大气科学专业系列英文图书

Practice Tutorial: Programming and Graphing for Atmospheric Sciences

Yang Chun
Chen Lin
Yu Yueyue

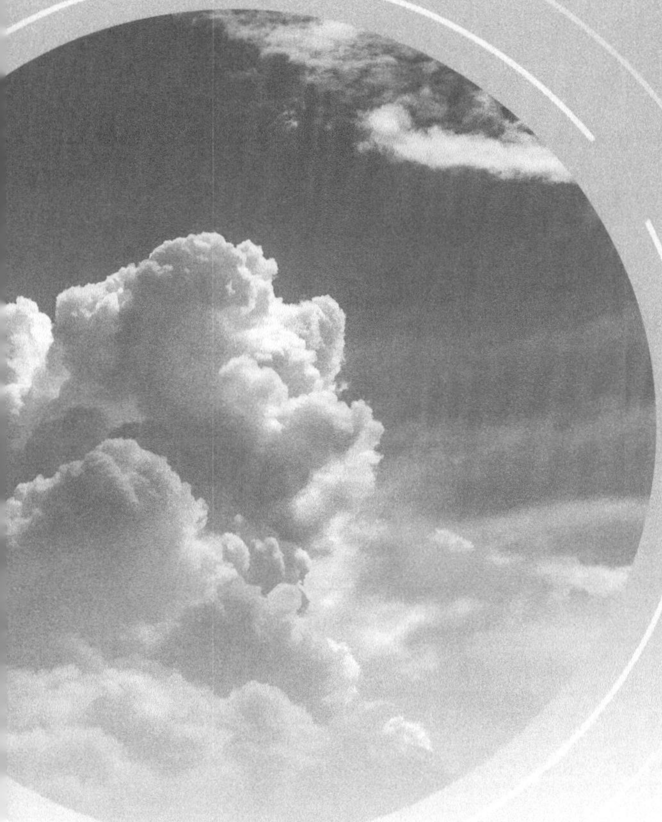

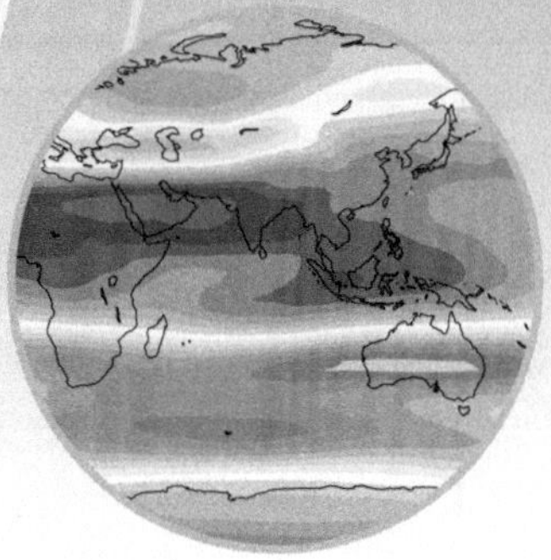

Practice Tutorial: Programming and Graphing for Atmospheric Sciences

This book is aimed at the practical application of Fortran and NCAR Command Language (NCL). It is to strive for complete theory, rich examples, detailed explanation and full practice. It is divided into 9 chapters, mainly including Introduction to Fortran and NCL, Fundamentals of Fortran Language, Selective and Repetitive Execution, Arrays, Subprograms, Files, as well as one-dimensional and two-dimensional graphic drawing techniques and composite analysis applications of NCL. To facilitate students' learning, each chapter firstly clarifies the key and difficult points of the knowledge, helping students to grasp the knowledge structure and deepen their understanding of the knowledge. By analyzing examples from shallow to deep level, students can deeply understand the application of these tools in their majors. Finally, exercises are provided to promote students' independent thinking and improve their hands-on abilities.

This book is suitable for beginners and advanced students of Fortran and NCL, and can serve as a reference book for learning two languages.

图书在版编目（ＣＩＰ）数据

大气科学编程与绘图实践教程 ＝ Practice Tutorial: Programming and Graphing for Atmospheric Sciences: 英文 / 杨春, 陈林, 虞越越编著. —— 北京：气象出版社, 2024.1
　　ISBN 978-7-5029-8103-7

　Ⅰ．①大… Ⅱ．①杨… ②陈… ③虞… Ⅲ．①软件工具—程序设计—应用—大气科学—教材—英文 Ⅳ．①P4

中国国家版本馆CIP数据核字（2023）第221281号

Practice Tutorial: Programming and Graphing for Atmospheric Sciences
By Yang Chun, Chen Lin, and Yu Yueyue
Responsible editor: Huang Haiyan

Copyright © 2024 by China Meteorological Press
Published by China Meteorological Press
No. 46, Zhongguancun Nandajie, Haidian District, Beijing 100081, China

http://www.qxcbs.com
E-mail: qxcbs@cma.gov.cn
Tel: 86-10-68407112

Printed in Beijing
First published in Jan. 2024

Preface

In meteorological operational services and research, it is necessary to use computer language and professional visualization language to process and analyze meteorological data, which is important for studying the complex changes of weather and climate systems. The programming and drawing ability are also important criteria for measuring the professional quality and ability of students who majored in meteorology, which can help lay a solid foundation for their future study and work.

On the basis of the module of "Meteorological Programming and Plotting", this book is aimed at the practical application of Fortran and NCAR Command Language (NCL), which are widely used in the field of meteorology. The content of this book is compiled combining the author's years' of teaching experience, and the related textbooks and materials worldwide. This book is to strive for complete theory, rich examples, detailed explanation and full practice.

This book is divided into 9 chapters, mainly including Introduction to Fortran and NCL, Fundamentals of Fortran Language, Selective and Repetitive Execution, Arrays, Subprograms, Files, as well as one-dimensional and two-dimensional graphic drawing techniques and composite analysis applications of NCL. To facilitate students' learning, each chapter firstly clarifies the key and difficult points of the knowledge, helping students to grasp the knowledge structure and deepen their understanding of the knowledge. By analyzing examples from shallow to deep level, deconstructing the practical application of language, focusing on majors, and analyzing classic meteorological application problems, students can deeply understand the application of these tools in their majors, and enhance their ability to analyze and solve meteorological problems. Finally, exercises are provided to promote students' independent thinking and improve their hands-on abilities.

The purpose of this book is to enable students to understand the basic syntax of Fortran and NCL languages through learning, master the debugging methods and steps of programs, proficiently use both languages to analyze and visualize meteorological data, effectively solve practical problems encountered in meteorological major, significantly improve students' computer application abilities, and lay the foundation for professional application and development.

The compilation of this book has been supported by Jiangsu Province International Talent Cultivation Brand Major Construction Project and Teaching Material Construction Fund Project of Nanjing University of Information Science & Technology.

Due to the limited proficiency of the authors and the urgency of time, there are inevitably some inappropriate aspects in the book. Readers and peer experts are kindly requested to criticize and correct them.

Authors

May 2023

Contents

Preface

Introduction ··· (1)

Chapter 1 Basic Fortran ·· (3)

Chapter 2 Controlling Execution ·· (11)

   2.1 Selective Execution ·· (11)

   2.2 Repetitive Execution ·· (24)

Chapter 3 Array ·· (38)

   3.1 One-dimensional array ·· (38)

   3.2 Multi-dimensional array ·· (41)

Chapter 4 Subprogram ·· (57)

   4.1 Function ··· (57)

   4.2 Subroutine ·· (63)

Chapter 5 File ·· (70)

Chapter 6 One-dimensional Plot ·· (74)

Chapter 7 Two-dimensional Plot ·· (83)

   7.1 Contours and shading ··· (83)

   7.2 Vector figure ·· (90)

   7.3 Overlay plot ··· (98)

Chapter 8 Synthesized Example ··· (107)

   8.1 Spectral analysis ·· (107)

   8.2 EOF analysis ··· (115)

   8.3 Composite analysis ·· (125)

   8.4 Regression and correlation analysis ·· (131)

Introduction

(1) Fortran Language

FORTRAN, FORmula TRANslation, the first "high-level" computer programming language, is extensively used in scientific, engineering computing. It is an old programming language with a history of 60 years, but still has a strong user base with scientific programmers in different organizations, such as meteorology, financial trading, and engineering simulations. Especially, most of the numerical models for meteorological and climatological use are written with Fortran language.

Fortran is a compiled language. The Fortran code must be compiled before running it on a computer. This is the part Fortran differs from interpreted languages such as Python, which executes the instructions directly at the cost of computing speed. Fortran has the advantages of high standardization, easy optimization and fast computing speed.

Fortran was developed by John Backus in 1954, released as Fortran I commercially in 1957. The first Fortran compiler was a milestone in the history of computing. FORTRAN II (1958), FORTRAN III (1958), FORTRAN IV (1961) was used to retain backward compatibility with earlier FORTRAN programs. Currently, the most commonly used versions include Fortran 77 (1978), Fortran 90 (1991), and Fortran 95 (1997).

For this course, Fortran 90 was selected for programing. It is free format on source code, not fixed format as Fortran 77. It added various sorts of threading, and direct array processing, and allowed other programming languages to evolve and compete with Fortran. The latest Fortran standard was released in 2018, bringing new features and keeping Fortran a relevant and highly performant language for contemporary scientific computing challenges. Fortran's advantages keep it continuously widely used in a number of specialized scientific communities.

(2) NCAR Command Language (NCL)

NCL stands for "NCAR Command Language", which is an interpreted language designed specifically for scientific data analysis and visualization. Developed by the National Center for Atmospheric Research (NCAR), NCL is portable, robust, free, and available for downloading on multiple platforms including Windows, macOS and Linux. Its official website is https://www.ncl.ucar.edu/. NCL provides a powerful set of tools, libraries, and modules that allow researchers to efficiently process diverse large datasets, perform complex statistical analyses, and to create publication-quality graphics, etc.

NCL has a complete data processing environment, which provides a one-stop service from reading data and processing to the final output. It supports various data formats such as NetCDF, GRIB2, HDF4/5, etc., making it useful across several disciplines including atmospheric science, oceanography, and climate research. It contains various commonly-used functions of modern programming languages, e. g., conditional statements, loops, array operations. NCL also includes many useful built-in functions and procedures for processing and manipulating data, such as statistical functions, interpolation, EOF analysis, spectral analysis, and so on. In addition to its abundant function library, NCL can call subroutines written by Fortran and C.

One of the most notable features of NCL is its excellent visualization capability. With built-in capabilities, a wide range of plot types is created, including contour plots, scatterplots, and maps among others along with well-designed color tables that are tailored to specific applications. On the NCL website, many example scripts and their corresponding graphics are available.

Therefore, NCL's unique combination of ease-of-use nature combined with powerful scripting ability makes it an essential tool for any scientist looking to derive insights from complex datasets or produce high-quality graphics for presentations or publications. It is widely used in scientific research and education, especially in meteorology, oceanography, and geophysics.

Chapter 1 Basic Fortran

A Fortran program is made of a collection of program units, such as main program, subroutine, function and modules. Each program must contain one main program and can contain other program units.

➢ **Key points**

(1) The general form of a main program

          PROGRAM program-name
          IMPLICIT NONE
            [! comments]
            [declaration statements]
            [executable statements]
          END [PROGRAM [program-name]]

• Contents in [ ] are optional command lines, and all the following examples are in the same fashion.

• All Fortran programs start with the keyword **PROGRAM** and end with the keyword **END** or **END PROGRAM**, or followed by the name of the program.

• The **IMPLICIT NONE** statement, which disables the default naming convention of variables, must be present at the start of every program unit.

• Comments in Fortran are started with the exclamation mark (!), and all characters after this are ignored by the compiler.

• *declaration statements* and *executable statements* are two basic parts of Fortran code.

• Fortran allows both uppercase and lowercase letters. Fortran is case-insensitive, except for string literals.

• If a statement is too long to fit on a line, it can be continued with an ampersand, &, at the end of this line.

• Indentation of code lines is a good practice for keeping a program readable and organized.

(2) Data type

Each executable statement performs a task that makes use of different types of data. Fortran provides five intrinsic data types: **INTEGER**, **REAL**, **COMPLEX**, **LOGICAL** and **CHARACTER**.

(3) Identifier

A Fortran identifier is a name used to identify a variable, constant, array, or any other

user-defined item. It must satisfy the following rules:
- It has no more than 31 characters.
- It can be made of letters, digits, or underscores, but the first character must be a letter.
- The identifiers are case insensitive.

For example, nuist and NUIST are the same identifier; nuist, nul_st are legal identifiers, but _nuist, nuis$t are illegal identifiers.

(4) Variable and variable declaration

The variable has a *name* and a *type*. The variable name is the identifier. There are five basic data types, i.e., **INTEGER**, **REAL**, **COMPLEX**, **LOGICAL** and **CHARACTER**.

The variables should be declared at the start of a program unit with a type declaration statement, as follows:

$$\text{type-specifier} \quad \text{var-list}$$

where the *type-specifier* is one of the five basic data types and *var-list* is a list of variable names separated with commas.

For the character variable, the length should be given in the declaration statement.

For example:

```
integer a, b
real c, d
character* 4 e, f
```

By such command lines, variables *a* and *b* are declared as integer type, and variable *c* and *d* are real type. Variables *e* and *f* are character type with a length of 4.

To initialize variables in the declaration part, the following statement can be used:

```
type-specifier:: var1= value1, var2= value2 ···
```

The separator (::) is required in the type declaration statement. For example:

```
integer:: a= - 5, b= 2
real:: c= 2.3, d= 3E+ 10
character* 3:: e= "all", f= "nuist"
```

(5) Constant and constant declaration
- A literal constant has a *value*, but no name (Table 1.1).

Table 1.1 Constants type and examples

Type	Example
integer	- 1, 0, 99
real	- 3.2, 0.0, 12.5, 3.2E+ 10
complex	(3.0, 5.0)
logical	.true.,.false.
character	"hello", "world_$"

- A named constant has a *value* and a *name*.

Named constants should be declared at the beginning of a program unit with the

PARAMETER attribute as

                `type-specifier, parameter::constant-name= value`

where *parameter* is the keywords for constant declaration. The double colon (::) is required. For example:

                `real, parameter::pi= 3.1415926535`

After assigning a name to a constant value, the constant can use the name throughout the program. The compiler would convert that name to its corresponding value. That name is not a variable.

(6) Operators and Expressions

Fortran has four types of operators: arithmetic, relational, logical, and character.

• Arithmetic Operators and Expressions (Table 1.2)

Table 1.2   Arithmetic operators and expression examples

Operator	Description	Example ($a=4$, $b=2$)
+	Addition Operator	a+b will give 6
-	Subtraction Operator	a-b will give 2
*	Multiplication Operator	a*b will give 8
/	Division Operator	a/b will give 2
**	Exponentiation Operator	a**b will give 16

The arithmetic expression is an expression with arithmetic operators.

The result type of an arithmetic expression is identical to that of the operands, if the operands are of the same type. If operands contain both **INTEGER** and **REAL** constants or variables, **INTEGER** operands are first converted to **REAL** type and then the operation is performed.

For example:

1/2 will give 0, not 0.5, because 1 and 2 are integers, the answer should be integer.

1/2.0 will give 0.5, because 2.0 is real, 1 will be converted to 1.0, the answer will be real too.

$\sqrt[3]{5}$ should be written as 5** (1.0/3.0) not 5** (1/3).

• Relational Operators and Expressions (Table 1.3)

Table 1.3   Relational operators and expression examples

Operator	Equivalent	Description	Example ($a=4$, $b=2$)
==	.eq.	equal to	a==b is. FALSE.
/=	.ne.	not equal to	a/=b is. TRUE.
>	.gt.	greater than	a>b is. TRUE.
>=	.ge.	greater than or equal to	a>=b is. TRUE.
<	.lt.	less than.	a<b is. FALSE.
<=	.le.	less than or equal to	a<=b is. FALSE.

Relational expression is conducted with two operands and one relational operator. The following rules are required:

① The two operands must be both arithmetic or be both character strings.

② If two operands are both arithmetic expressions, they are converted into the same type before conducting relational operations, and the conversion method is the same as arithmetic expressions.

③ If two operands are character expressions, they are converted into same-length strings before conducting relational operations.

④ The result of a relational expression is a **LOGICAL** value.

⑤ All relational operators are the same priority, which are lower than those of arithmetical operators.

For example:

12>34 Result:. FALSE

(4+5*2).LE.10 Result:. FALSE.

"banana"< = "apple" Result:. FALSE. The ASCII of the character "a" is 97, while the ASCII of the character "b" is 98.

• Logical Operators and Expressions (Table 1.4)

Table 1.4 Logical operators and expression examples

Operator	Description	Example (a is. TRUE. , b is. FALSE. , c is. TRUE.)
.and.	Logical AND Operator	a.and.b is. FALSE.
.or.	Logical OR Operator	a.or.b is. TRUE.
.not.	Logical NOT Operator	.not.a is. FALSE. ; .not.b is. TRUE.
.eqv.	Logical EQUIVALENT Operator	a.eqv.b is. FALSE. ; a.eqv.c is. TRUE.
.neqv.	Logical NON-EQUIVALENT Operator	a.neqv.b is. TRUE. a.neqv.c is. FASLSE.

Logical expression can only be used with logical operators and operands, whose results are logical values (. **TRUE.** and. **FALSE.**). All logical operators' priorities are lower than arithmetic and relational operators. The result of a logical expression is a **LOGICAL** value.

(7) Assignment statement

The form of assignment statement is

variable= expression

The result of right *expression* is assigned to the left *variable*. If the type of *expression* result is different from the *variable*, the result type should be converted to the type of *variable*.

The initial value can be assigned to the variable with declaration statements or **DATA** statement.

$$\text{DATA variable-list/value-list/}$$

DATA is the key word, the *variable-list* is a list of variable names separated with commas, the *value-list* between two '/' is the list of initial values for variables. Usually, the **DATA** statement is placed between declaration statements and executable statements.

(8) List-Directed Input-Output (I/O)

• List-directed input is carried out with the **READ** statements. The **READ** statement can be used to read input values into a set of variables with the keyboard. The form of **READ** statement is:

$$\text{read(* ,*) io-list}$$
$$\text{read* , io-list}$$

where the first asterisk (*) means the input comes from the keyboard; the second asterisk (*) means the input format is default format. *io-list* is a list of variables separated with commas.

• List-directed output is carried out with the **WRITE** statements or **PRINT** statements to display the results on the screen. The form of output statement is:

$$\text{write(* ,*) io-list}$$
$$\text{print* , io-list}$$

The first asterisk (*) means the output display on the screen. The second asterisk (*) means the output format is default format.

(9) Formatted Input-Output (I/O)

The purpose of formatted I/O is displaying data in a certain format. The form is:

$$\text{read(* ,"(fmt)") io-list}$$
$$\text{write(* ,"(fmt)") io-list}$$

The *fmt* is the format specification containing a list of edit descriptors to define the way in which formatted data is displayed (Table 1.5).

Table 1.5 Edit descriptors and examples

Edit Descriptor	Description	Example
Iw[.m]	Reading/writing INTEGERs	write(* ,"(i3)") a
Fw.d	Reading/writing REALs in decimal form	write(* ,"(2f3.1)") a, b
Ew.d	Reading/writing REALs in exponential form	write(* ,"(e10.3)") a
Aw	Reading/writing CHARACTERs	write(* ,"(a10)") a
nX	Output space	write(* ,"(10x, i3)") a
/	Insert blank lines	write(* ,"(/10x, i3)") a

w is the number of positions to be used; m is the minimum number of positions to be used; d is the number of decimal places; n is the number of desired space.

For example:

character* 3:: a= "nuist"   ! the character length declaration is 3, so a is

"nui"
```
   write(* ,"(a5)") a
```
The output format for a is a5, the length is 5, but a's length is 3, thus filling the space before the value of a.

The result on the screen is:

□□nui

> **Examples**

Example 1-1:

This is a simple complete Fortran program.
```
1    program ex1_1
2    implicit none
3      real st1,st2,stave
4      st1= 8.5
5      st2= 9.0
6      stave= (st1+ st2)/2.0
7      print * , 'stave=',stave
8    end
```

Decode:

Line 1—2 are the two header statements of a Fortran main program with the keyword **PROGRAM**, the program name $ex1\_1$, and **IMPLICIT NONE** statement.

Line 3 is the declaration statement, declaring 3 real variables.

Line 4—6 are the assignment statements, assigning two real constants to variables $st1$ and $st2$, and giving the result of an arithmetic expression to variable $stave$.

Line 7 is the output with **PRINT** statement, and the results will be shown on the screen directly.

Line 8 is the end of this program with **END** statement.

Results:

```
stave=    8.750000
Press any key to continue . . .
```

Example 1-2:
```
1    program ex1_2
2    implicit none
3      ! declaring variables
4      integer a
5      real b
6      complex c
7      logical d
8      character(len= 80) e
9      ! assigning values
```

```
10      a= 200
11      b= 1.6
12      c= (3.0, 5.0)
13      d = .true.
14      e= "Program tutorials of Fortran"
15      write(* ,"(i3)") a
16      write(* ,"(f3.1)") b
17      write(* ,"(f3.1,1x,f3.1)") c
18      write(* ,"(i3)") d
19      write(* ,"(a20)") e
20      end program
```

Decode:

Line 1-2 are two header statements of a Fortran main program.

Line 3 and Line 9 are comment lines which start with "!".

Line 4-8 are the declaration statements, declaring 5 variables with 5 different types.

Line 10-14 are the assignment statements, assigning different values to different variables.

Line 15-19 are the formatted output statements, and the results will be shown on the screen with the given format for different type variables.

Line 20 is the end of this program.

Results:

```
200
1.6
3.0 5.0
 -1
Program tutorials of
Press any key to continue . . .
```

Example 1-3:

Write a program to compute and display the geometric mean of three **REAL** variables input from keyboard. The geometric mean is $\sqrt[3]{a \times b \times c}$.

Decode:

According to the question, three real variables should be declared to save the values input with keyboard. Because the geometric mean of them is also real type, another real variable should be defined for the result. The input and output statements are list-directed I/O statements.

Code:

```
1       program ex1_3
2       implicit none
3       real a, b, c, gmean
4       read* , a, b, c
5       gmean= (a* b* c)* * (1.0/3.0)
```

```
6     print* , gmean
7     end
```

Line 1—2 are the head lines of program.

Line 3 is the declaration statement to declare 4 real variables.

Line 4 is the read statement to input three real variables from keyboard.

Line 5 is to calculate the geometric mean of variable a, b and c, and assign the result to *gmean*. The cube root is computed by " * * (1.0/3.0) ", where 1 or/and 3 should be real data type; if both of them are integer types, the cube root result will be 1.

Line 6 is to display the value of *gmean* on screen with list-directed output statement.

Line 7 is the end of program.

Results:

Input 1.2, 3.6, 9.0, the output result on screen will be as follows.

```
1.2, 3.6, 9.0
   3.387730
Press any key to continue . . .
```

➤ **Practices**

(1) Write a program to output the following strings on the screen.

　　　　　　　Hello everyone

　　　　　　　This is "programming and graphing for meteorology"

(2) Write a program to calculate the area of a circle. Declare π as a constant. Input the radius from keyboard, and output the result on the screen with the format f4.1.

Chapter 2 Controlling Execution

2.1 Selective Execution

Logical judgments are made according to the given conditions, based on the results of judgments, the compiler decides which operation should be performed—*selective execution*.

2.1.1 IF statement

➤ **Key points**

(1) Single branch **IF** statement

```
IF (condition) THEN
    statements
END IF
```

where *statements* are a sequence of executable statements, and *condition* is a logical expression.

The execution of this form of **IF** statement goes as follows:

- Firstly, the *condition* is evaluated, yielding a logical value.
- If the result is. **TRUE.**, the statements in *statements* are executed.
- If the result is. **FALSE.**, the statement following the **END IF** is executed. In other words, if the *condition* is. **FALSE.**, there is no action taken.
- Indentation of code lines is a good practice for keeping a program readable.

(2) Double branch **IF** statement

```
IF (condition) THEN
    statements-1
ELSE
    statements-2
END IF
```

where *statements*-1 and *statements*-2 are executable statements, and *condition* is a logical expression.

The execution of this form of **IF** statement goes as follows:

- The *condition* is evaluated, yielding a logical value.
- If the result is. **TRUE.**, the statements in *statement*-1 are executed.
- If the result is. **FALSE.**, the *statement*-2 following the **ELSE** is executed.

(3) Multi-branch **IF** statement

```
IF (condition-1) THEN
    statements-1
ELSE IF (condition-2) THEN
    statements-2
ELSE IF (condition-3) THEN
    statements-3
...
ELSE IF (condition-n) THEN
    statements-n
[ELSE
    statements-else]
END IF
```

where *statements*-1, ⋯, *statements*-n are executable statements, and *condition*-1, ⋯, *condition*-n are logical expressions, and [⋯] means the content in [ ] is optional.

The execution of this form of **IF** statement goes as follows:

• The *condition*-1 is evaluated, yielding a logical value.

• If the result is. **TRUE**. , *statements*-1 is executed.

• If the result is. **FALSE**. , Fortran evaluates *condition*-2 and executes *statements*-2 following **ELSE IF**. Otherwise, Fortran continues to evaluate the next *condition*.

• If all *conditions* are. **FALSE**. , and if **ELSE** is there, *statements-else* is executed; otherwise, Fortran executes the statement after the **END IF**.

(4) Logical **IF** statement

```
IF (condition) statement
```

where *statement* is an executable statement, and *condition* is a logical expression. Only one statement is here.

The execution of this form of **IF** statement goes as follows:

• The *condition* is evaluated, yielding a logical value.

• If the result is. **TRUE**. , the *statement* is executed.

• If the result is. **FALSE**. , the statement following the logical **IF** is executed.

(5) Nested **IF** statement

```
IF (condition) THEN
    statements
    IF (condition) THEN
        statements
    ELSE
        statements
    END IF
    statements
```

```
                ELSE
                    statements
                    IF (condition) THEN
                        statements
                    END IF
                    statements
                END IF
```
where *statements* are an executable statement, and *conditions* are logical expression. One or more **IF** statements with different forms can be contained following the **THEN** or the **ELSE**.

> **Examples**

Example 2-1-1:

Compute y with the following formula:
$$y = 10.0/x$$

Input an integer with keyboard into variable x. If x is non-zero, calculate y and output the value on the screen.

Decode:

For this question, the value of x is an integer, so x should be declared as integer type. Based on the formula, y is a real number, so y should be declared as real type.

Only when x is non-zero, the code can be written with single branch IF statement or logical IF statement. If the condition's result is true, y will be computed and outputted on the screen with list-directed output format.

Code:

• Programming with single branch IF statement

```
1    program ex2_1_1
2    implicit none
3      integer x
4      real y
5      read* , x
6      ! single branch IF statement
7      if(x/= 0)then
8         y= 10.0/x
9         print* ,"x= ",x, "  y= ",y
10     end if
11   end
```

Line 1—2 declare the beginning of main program.

Line 3 declares x as an integer variable.

Line 4 declares y as a real variable.

Line 5 is to list-directed input value to x.

Line 6 is a comment line.

Line 7 judges the condition $x \neq 0$.

Line 8—9 are the operation steps for the case when the judgement is .TRUE., calculate y and output it with list-directed format.

Line 10 ends the IF statement.

Line 11 is the end of this main program.

Results:

If input 9 to x, the output result on screen will be as follows.

```
9
 x=           9    y=    1.111111
Press any key to continue . . .
```

• Programming with logical IF statement

```
1    program ex2_1_1
2    implicit none
3    integer x
4    real y
5    read* , x
6    ! logical IF statement
7    if(x/= 0)y= 10.0/x
8    print* ,"x= ",x, "   y= ",y
9    end
```

Line 7 is to judge if the condition $x \neq 0$; if the result is .TRUE., calculate y with $y = 10.0/x$.

Results:

Input 10 to x, the output result on screen will be as follows.

```
10
 x=          10    y=    1.000000
Press any key to continue . . .
```

Example 2-1-2:

Compute y with the following formula:

$$y = \begin{cases} 10.0/x & x > 10 \\ 2x^2 & x \leq 10 \end{cases}$$

Read an integer into variable x. Calculate y and output the value on the screen.

Decode:

As in Example 2-1-1, x and y should be declared as integer type and real type, respectively.

There are two conditions: the condition-1 is $x > 10$, and the condition-2 is $x \leq 10$. The code can be written with double branch IF statement. If the condition-1's result is true, operate $y = 10.0/x$. Otherwise, $y = 2x^2$. Finally, output the value of y on the screen with list-directed output format.

Code:

```
1   program ex2_1_2
2   implicit none
3     integer x
4     real y
5     read* , x
6     !   double branches IF statement
7     if(x> 10)then
8         y= 10.0/x
9     else
10        y= 2* x* * 2
11    end if
12    print* ,"x= ",x, "  y= ",y
13  end
```

Line 1—2 declare the beginning of main program.

Line 3 declares x as an integer variable.

Line 4 declares y as a real variable.

Line 5 is to list-directed input value to x.

Line 6 is a comment line.

Line 7 judges the condition-1 $x>10$.

Line 8: if condition-1's result is. TRUE. , execute this line.

Line 9 — 10: if condition-1's result is. FALSE. , namely $x \leqslant 10$, execute ELSE statement.

Line 11 is the end of IF statement.

Line 12 is to output y with list-directed output statement.

Line 13 is the end of this main program.

Results:

Input 6 to x, the output result on screen will be as follows.

```
6
 x=           6   y=       72.00000
Press any key to continue . . .
```

Example 2-1-3:

Compute y with the following formula:

$$y=\begin{cases} 10.0/x & x>10 \\ 2x^2 & 5<x\leqslant 10 \\ 0 & x\leqslant 5 \end{cases}$$

Input an integer into variable x. Calculate y and output the value on the screen.

Decode:

As the above examples, x and y should be declared as integer type and real type, respectively.

There are three conditions: the condition-1 is $x>10$, the condition-2 is $5<x\leqslant 10$, and the condition-3 is $x\leqslant 5$. The code can be written with multi-branch IF statement or nested IF statement. If the condition-1's result is true, $y=10.0/x$. If the condition-1's result is false and condition-2's result is true, $y=2x^2$. If the result of condition-2 is also false, which means x is less than or equal to 5, $y=0$. Finally, output the value of y on the screen with list-directed output statement.

Code:
- Programming with multi-branch IF statement

```
1    program ex2_1_3
2    implicit none
3      integer x
4      real y
5      read* , x
6      ! multi-branch IF statement
7      if(x> 10)then
8         y= 10.0/x
9      else if (5< x< = 10) then
10        y= 2* x* * 2
11     else
12        y= 0
13     end if
14     print* ,"x= ",x, "   y= ",y
15   end
```

Line 1—2 declare the beginning of the main program.
Line 3 declares x as an integer variable.
Line 4 declares y as a real variable.
Line 5 is to list-directed input value to x.
Line 6 is the comment line.
Line 7 judges the condition-1 $x>10$.
Line 8: if condition-1's result is. TRUE. , execute this line.
Line 9—10: if condition-1's result is. FALSE. , and condition-2's result is. TRUE. , execute ELSE IF statement.
Line 11—12: if condition-2's result is also. FALSE. , execute ELSE statement.
Line 13 is the end of IF statement.
Line 14 outputs y with list-directed format.
Line 15 is the end of this main program.

Results:
Input 6 to x, the output result on screen will be as follows.

```
x=           6    y=    72.00000
Press any key to continue . . .
```

- Programming with nested IF statement

```
1    program ex2_1_3
2    implicit none
3      integer x
4      real    y
5      read* , x
6      !   nested IF statement
7      if(x> 10)then
8         y= 10.0/x
9      else
10        if (x> 5) then
11           y= 2* x* * 2
12        else
13           y= 0
14        end if
15     end if
16     print* ,"y= ", y
17   end
```

Line 7 is the outer IF statement with condition-1.

Line 8: if condition-1's result is .TRUE. , execute this line.

Line 9: if the condition-1's result is .FALSE. , which means $x \leqslant 10$, execute ELSE statement.

Line 10 declares the inner IF statement with condition-2, $x>5$, which means $5<x \leqslant 10$.

Line 11: if condition-2's result is .TRUE. , execute this line.

Line 12—13: if condition-2's result is .FALSE. , which means $x \leqslant 5$, execute the inner ELSE statement.

Line 14 is the end of inner IF statement.

Line 15 is the end of outer IF statement.

Results:

Input 6 to x, the output result on screen will be as follows.

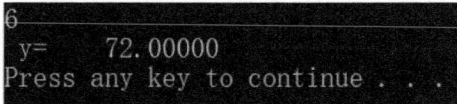

Example 2-1-4:

For meteorology, the wind can be divided into two components, U and V.

If $U>0$ and $V>0$, it is southwest wind. If $U>0$ and $V<0$, it is northwest wind. If $U>0$

and $V=0$, it is west wind.

If $U<0$ and $V>0$, it is southeast wind. If $U<0$ and $V<0$, it is northeast wind. If $U<0$ and $V=0$, it is east wind.

If $U=0$ and $V>0$, it is south wind. If $U=0$ and $V<0$, it is north wind. If $U=0$ and $V=0$, it is no wind.

Input U and V, output the wind direction with list-directed input/output statements.

Decode:

For this question, two variables should be declared to save the value of U and V. The wind direction results can be output on the screen directly as a string.

There are 9 cases, which can be divided into 3 outer-conditions as $U>0$, $U<0$, and $U=0$. Each U condition can be spread into 3 inner-conditions as $V>0$, $V<0$, and $V=0$. Therefore, the code can be written with multi-branch IF statement for outer-conditions and nested IF statement for inner-conditions.

Code:

```
1    program ex2_1_4
2    implicit none
3      real u, v
4      character* 30 wind_dir
5      read* , u, v
6      if (u> 0)then
7        if (v> 0) then
8           wind_dir= "southwest wind"
9        else if (v< 0) then
10          wind_dir= "northwest wind"
11       else
12          wind_dir= "west wind"
13       end if
14     else if (u< 0)then
15       if (v> 0) then
16          wind_dir= "southeast wind"
17       else if (v< 0) then
18          wind_dir= "northeast wind"
19       else
20          wind_dir= "east wind"
21       end if
22     else
23       if (v> 0) then
24          wind_dir= "south wind"
25       else if (v< 0) then
```

```
26           wind_dir= "north wind"
27         else
28           wind_dir= "no wind"
29         end if
30       end if
31       print* , "it is ", wind_dir
32     end program
```

Line 3—4 are variable declaration statements for u, v and the wind direction $wind\_dir$.

Line 5 inputs the value of u, v with the keyboard.

Line 6/14/22/30 are the outer IF statements with IF-ELSE IF-ELSE-END IF for u conditions.

Line 7—13 are the first inner IF statements for $u>0$ with $v>0$, $v<0$ and $v=0$.

Line 15—21 are the second inner IF statements for $u<0$ with $v>0$, $v<0$ and $v=0$.

Line 23—29 are the third inner IF statements for $u=0$ with $v>0$, $v<0$ and $v=0$.

Results:

Input 5 to u, and -9 to v, and the output result on screen will be as follows.

```
5, -9
it is northwest wind
Press any key to continue . . .
```

Example 2-1-5:

Input 3 integers, and reorder them from small to large.

Decode:

The code can be written with "the bubbling sorting method" to reorder the numbers. The principle is to compare two adjacent numbers and exchange the numbers with large values to the right.

For example, there are four numbers, 6, 5, 4, 3. The flow chat is:

number	6	5	4	3	
First pass	5	6	4	3	Compare the first two numbers, and swap since 6>5
	5	4	6	3	Swap since 6>4
	5	4	3	6	Swap since 6>3
Second pass	4	5	3	6	Swap since 5>4
	4	3	5	6	Swap since 5>3
Third pass	3	4	5	6	Swap since 4>3

Code:

```
1   program ex2_1_5
2   implicit none
3     integer a, b, c, min
4     read* , a, b, c
```

```
5      if (a> b) then
6         min= b
7         b= a
8         a= min
9      end if
10     if (b> c) then
11        min= c
12        c= b
13        b= min
14     end if
15     if (a> b) then
16        min= b
17        b= a
18        a= min
19     end if
20     print* , "from small to large: ", a, b, c
21  end program
```

Line 3 declares *a*, *b*, *c*, *min* as integer variables. Variable *min* is used to save the minimum between two comparable numbers.

Line 5—9 are the first pass, if $a > b$, *b* is the minimum. The value of *b* is assigned to *min*, the value of *a* is assigned to *b*, then update *a* as the minimum. The purpose is to swap *a* and *b*.

Line 10—14 indicate that if $b > c$, swap *b* and *c*.

Line 15—19 are the second pass, if $a > b$, swap *a* and *b*.

Line 20 is to output the result.

Results:

Input 1, 45, 34, the output result on screen will be as follows.

```
1 45 34
 from small to large:         1         34         45
Press any key to continue . . .
```

> **Practices**

Write the code with IF statements for the following questions.

(1) Input one number, determine whether it is even or odd.

(2) Tropical cyclone is one of the most destructive weather systems. According to its maximum wind speed, it can be divided into different levels as the following table.

Maximum wind speed/(m/s)	Tropical cyclone grade
10.8—17.1	Tropical depression
17.2—24.4	Tropical storm

续表

24.5—32.6	Severe tropical storm
32.7—41.4	Typhoon
41.5—50.9	Severe typhoon
≥51.0	Super typhoon

Input the maximum wind speed, and output the grade of cyclone.

(3) Given a quadratic equation $ax^2+bx+c=0$, where $a\neq 0$, its roots are computed as follows:

$$x=\frac{-b\pm\sqrt{b^2-4ac}}{2a}$$

If $b^2-4ac\geqslant 0$, write the code to read a, b, and c, calculate and display the roots on screen.

(4) Input the ID of 10 students and their test scores, and output the students' ID, whose scores are larger than average scores.

ID	01	02	03	04	05	06	07	08	09	10
Score	73	76.5	89	64	90	91	79	85.5	94	87

(5) Input 5 integers, and output the maximum.

2.1.2 SELECT CASE statement

> **Key points**

(1) The syntactic form of **SELECT CASE** statement

```
SELECT CASE (selector)
    CASE (list-1)
        statements-1
    CASE (list-2)
        statements-2
    CASE (list-3)
        statements-3
        ············
    CASE (list-n)
        statements-n
    [CASE DEFAULT
        statements-DEFAULT]
END SELECT
```

where *selector* is an expression, whose result is of type **INTEGER**, **CHARACTER**, or **LOGICAL** (i.e., no **REAL** type can be used for the selector), and *statements*-1, *statements*-2, *statements*-3,..., *statements-n*, and *statements-DEFAULT* are sequences of executable statements. The label lists *list*-1, *list*-2, *list*-3,..., and *list-n*, called case labels, can be written as a list label, separated by commas. Each label can be described as an extent specifier for substrings, such as

$$value$$
$$value1:value2$$
$$value1:$$
$$:value2$$

where *value*, *value*1, and *value*2 are constants or **PARAMETERS** of the same type as *selector*.

The rule of executing the **SELECT CASE** statement goes as follows:

• The *selector* expression is evaluated.

• If the result is in *list-i*, then *statements-i* is executed, followed by the statement following **END SELECT**.

• If the result is not in any one of the *lists*, there are two possibilities:

If **CASE DEFAULT** is there, then *statements-DEFAULT* is executed, followed by the statement following **END SELECT**.

If there is no **CASE DEFAULT**, the statement following **END SELECT** is executed.

➢ **Examples**

Example 2-1-6:

As question (2) in 2.1.1. Write a program to read the wind speed, and output the grade of cyclone with SELECT CASE statement.

Decode:

For this question, one variable for wind speed should be declared. And, it also can be chosen as the *selector*. Mostly, wind speed is a real variable, so it should be converted to integer type first. The different wind speed ranges are the *lists* in CASE statement. If the wind speed inputted is in one range, the responding cyclone grade will be outputted.

Code:

```
1    program ex2_1_6
2    implicit none
3    real spd
4    read(* ,"(f4.1)") spd
5    spd= spd* 10
6    select case (int(spd))
7       case(108:171)
8          print* ," Tropical depression"
9       case(172:244)
10         print* ," Tropical storm"
```

```
11      case(245:326)
12         print* ,"Severe tropical storm"
13      case(327:414)
14         print* ," Typhoon"
15      case(415:509)
16         print* ," Severe typhoon"
17      case(510:)
18         print* ," Super typhoon"
19   end select
20  end
```

Line 3 declares *spd* as a real variable for wind speed.

Line 4 inputs *spd* with the given input format, f4.1, with one decimal place.

Line 5 converts the decimal part of wind speed by multiplying with 10.

Line 6 converts *spd* from real type to integer type used as the selector.

Line 7—18 contain different cases that correspond to different wind speed ranges, and output different statements.

Line 19 is the end of SELECT CASE statement.

Results:

Input 12.3, the output result on screen will be as follows.

```
12.3
 Tropical depression
Press any key to continue . . .
```

Example 2-1-7:

Write a program to read a student's score, output the grade of this student and the number of different grades with SELECT CASE statement.

Decode:

A variable should be used to save the students' score, which is the selector for this question. The different score ranges are the *lists* in CASE statement. If the score inputted is in one range, the responding grade will be outputted. Or, the grade result is assigned to a character variable, and finally, output the grade variable.

Code:

```
1   program ex2_1_7
2   implicit none
3     real score
4     character* 10 grade
5     read(* ,"(f4.1)")score
6     select case (int(score))
7       case(:59)
8          grade= "Poor"
```

```
 9        case(60:79)
10            grade= "Average"
11        case(80:89)
12            grade= "Good"
13        case(90:)
14            grade= "Excellent"
15     end select
16     print* ,grade
17  end program
```

Line 4 declares a character variable *grade* to save the grade result.
Line 6 converts the real *score* to integer *score* used as the selector.
Line 8/10/12/14 assign the result to variable *grade*.
Line 16 is to print out the result.

Results:

Input 59, the output result on screen will be as follows.

```
59
 Poor
Press any key to continue . . .
```

➢ Practices

(1) Use *class* as a selector. If *class* is 1, "Freshman" is displayed; if *class* is 2, "Sophomore" is displayed; if *class* is 3, "Junior" is displayed; if *class* is 4, "Senior" is displayed; and if *class* is none of the above, "I don't know" is displayed.

(2) Input a letter with the keyboard, and determine if it is an upper case letter or a lower case letter.

2.2 Repetitive Execution

2.2.1 DO statements

➢ Key Points

(1) The syntactic form of loops with **DO** statement

```
           DO loop-variable= start, end[, step]
              statements
           END DO
```

where *loop-variable* is an **INTEGER** variable (can be **REAL** type for Fortran 90). *start* and *end* are the initial value and final value of *loop-variable*. The *step* is the increment, which cannot be zero. It is assumed to have the value 1 if omitted. *statements* are sequences of executable statements.

The execution of **DO** statement goes as follows:

- Before the **DO**-loop starts, the values of *start*, *end* and *step* are computed and converted to the same type of *loop-variable*.
- Count the loop number R. the formula is: $R = \text{MAX}(\text{INT}((end-start+step)/step), 0)$.
- If R isn't 0, the *loop-variable* receives the value of *start*.
- When *step* is positive: if $R>0$ or the value of *loop-variable* is less than or equal to the value of *end*, the *statements* are executed. Then, R is reduced by 1 and the value of *step* is added to *loop-variable*. Go back and compare the values of *loop-variable* and *end*. If $R \leqslant 0$ or the *loop-variable* is greater than the value of *end*, the **DO**-loop completes and the statement following **END DO** is executed.
- When *step* is negative: if $R>0$ or the value of *loop-variable* is greater than or equal to the value of *end*, the *statements* are executed. Then, R is reduced by 1 and the value of *step* is added to *loop-variable*. Go back and compare the values of *loop-variable* and *end*. If $R \leqslant 0$ or the value of *loop-variable* is less than the value of *end*, the **DO**-loop completes and the statement following **END DO** is executed.
- Indentation of code lines is a good practice for keeping a program readable.

(2) Nested DO loops

The same as nested IF statement, a **DO** loop can contain other loops in its statements. The general form is:

```
DO loop-variable-1= start-1, end-1, step-1
    statements-1
    DO  loop-variable-2= start-2, end-2, step-2
        statements-2
        ......
    END DO
    statements-3
END DO
```

Each outer **DO** loop starts with *statements*-1. When the control reaches the inner **DO** loop, *statements*-2 is executed until some conditions of the inner **DO** bring the control out of it. Then, *statements*-3 is executed and starts the next outer **DO** loop.

➤ **Examples**

Example 2-2-1:

Given an integer *n*, write a program to calculate the factorial of *n* and output factorial with formatted output statements.

Decode:

The factorials of *n* is $n! = 1 \times 2 \times 3 \times 4 \times \cdots \times n$. A DO loop is selected with *n* as the end of loop variable. A variable should be declared to save the factorial of *n*. It should be assigned an initial value 1 first.

Code:

```
1    program ex2_2_1
```

```
2     implicit none
3     integer i, n, mu
4     read* ,n
5     mu= 1
6     do i= 1,n
7        mu= mu* i
8     end do
9     write(* ,"(a20,i3,a4,i3)") "The factorials of ", n, " is ", mu
10    end
```

Line 3 declares *i* as the loop variable, *n* as the integer, and *mu* as the factorial.
Line 4 inputs the value of *n* with a list-directed input statement.
Line 5 sets the initial value 1 to *mu*.
Line 6—8 are a DO loop with cycling from 1 to *n* to calculate the factorial.
Line 9 outputs the factorial with format i3.

Results:

Input 5 to n, and the output result on screen will be as follows.

```
5
     The factorials of   5 is 120
Press any key to continue . . .
```

Example 2-2-2:

Given 10 integer input values, write a program to count the number of even and odd values and compute their sum.

Decode:

The integer can be read with a DO loop, in which the end of loop variable is 10. 5 variables will be used to save the integer values, loop variable, the number of even or odd, and the sum of 10 integer values. The classic sum structure should be applied to calculate the number of even and odd values and the sum. The variable used as the number of even/odd and the sum should be initialized to 0 first.

Code:

```
1     program ex2_2_2
2     implicit none
3     integer a, i, nume, numo, summ
4     nume= 0
5     numo= 0
6     summ= 0
7     do i= 1,10
8        read* ,a
9        if(mod(a,2)= = 0) nume= nume+ 1
10       if(mod(a,2)/= 0) numo= numo+ 1
```

```
11 !  it also can be written as
12 !      if (mod(a,2)= = 0)then
13 !          nume= nume+ 1
14 !      else
15 !          numo= numo+ 1
16 !      end if
17        summ= summ+ a
18    end do
19    print* , "the number of even is: ", nume
20    print* , "the number of odd is: ", numo
21    print* , "the summation is: ", summ
22    end
```

Line 3 declares 5 integer variables.

Line 4—6 assign the initial value 0 to the number of even/odd values and their sum.

Line 7—18 are a DO loop with the number of integers as the loop variable.

Line 8 inputs an integer to variable a with a list-directed input statement in each loop.

Line 9 is a logical IF statement. The condition is a logical expression with MOD function. If the result is. TURE. , the integer is even. The number of even is updated by adding 1.

Line 10 is a logical IF statement. If the result is. TURE. , the integer is odd. The number of odd is updated by adding 1.

Line 11 — 16 show that two logical IF statements can be replaced with a single IF statement with two branches to determine whether the integer is even or odd.

Line 17 indicates that in each DO loop, the sum is updated by adding the new integer input in Line 8. In the last loop, *summ* is the sum of 10 integers.

Results:

Input 4, 5, 3, 4, 6, 7, 8, 2, 4, 1, the output result on screen will be as follows.

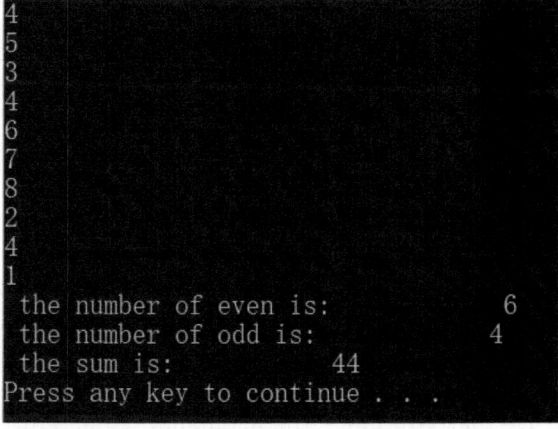

Example 2-2-3:

Given an integer n, calculate 1, 1+2, 1+2+3, 1+2+3+4, ⋯, 1+2+3+⋯+n and output them.

Decode:

The sums can be calculated with an outer DO loop with loop variable ranging from 1 to n. An inner DO loop can be configured to compute each sum with loop variable ranging from 1 to the current outer loop variable. For each outer loop, the sum should be initialized to 0 before the inner loop is executed.

Code:

```
1    program ex2_2_3
2    implicit none
3      integer i, j, summ, n
4      read* ,n
5      do i= 1,n
6         summ= 0
7         do j= 1,i
8            summ= summ+ j
9         end do
10        print* ,"the summation is ", summ
11     end do
12   end
```

Line 5—11 are an outer DO loop with loop variable *i* ranging from 1 to *n*.

Line 7—9 are an inner DO loop with loop variable *j* ranging from 1 to the current outer loop variable's value *i*.

Line 6 indicates that for each outer loop, *summ* should be initialized to 0 first.

Line 8 computes the *summ* by adding the inner loop variable values.

Results:

Input 7, the output result on screen will be as follows.

```
7
the summation is              1
the summation is              3
the summation is              6
the summation is             10
the summation is             15
the summation is             21
the summation is             28
Press any key to continue . . .
```

Example 2-2-4:

The temperatures of a week are input circularly to judge the maximum temperature and calculate the average temperature of the week. The daily temperature in Beijing from April 1 to 7, 2023 are as follows(unit:℃): 24.0, 21.0, 14.0, 13.0, 17.0, 19.0, 21.0.

Decode:

Input the temperature data with a DO loop with days as the loop variable extent from 1 to 7. To calculate the average, firstly, the sum of 7 temperature data should be computed. Initialize the sum to 0 before cumulating the calculation. Similarly, a real variable should be declared to save the maximum temperature with an initial value of 0. Compare the temperature data input in each loop with the current maximum temperature, if the result is. TRUE. , update the maximum temperature with the new data.

Code:

```
1    program ex2_2_4
2    implicit none
3      real::t, tmax= 0.0,sum= 0.0,tave
4      integer i
5      do i= 1,7
6        read * , t
7        sum= t+ sum
8        if (t> = tmax) then
9          tmax= t
10       end if
11     end do
12     tave= sum/7.0
13     print * , 'tmax= ', tmax
14     print * , 'tave= ', tave
15   end
```

Line 3 declares 4 real variables. *tmax* and *sum* are assigned the initial value of 0.0. Please pay more attention to the double colon (::) for setting the initial value in the variable declaration part.

Line 5—11 are the DO loop with loop variable *i* from 1 to 7.

Line 7 indicates that *sum* is updated by adding the input *t*.

Line 8—10 are the nested IF statements with $t \geq tmax$ as condition.

Line 9 indicates that if the condition is. TURE. , *tmax* is updated with *t*. If it is FLASE, *tmax* isn't changed.

Line 12 calculates the average temperature.

Results:

Input 24.0, 21.0, 14.0, 13.0, 17.0, 19.0, 21.0, and the output result on screen will be as follows.

```
24.0
21.0
14.0
13.0
17.0
19.0
21.0
 tmax=    24.00000
 tave=    18.42857
Press any key to continue . . .
```

➢ Practices

(1) Please find out all the "Armstrong number" from 100 to 999, whose value is equal to the sum of the cube of its digits, such as $153=1^3+5^3+3^3$.

(2) Please write the code to compute the following formula with given n and x.

$$e^x = 1+x+\frac{x^2}{2!}+\frac{x^3}{3!}+\cdots+\frac{x^n}{n!}$$

(3) Calculate the first 20 terms of Fibonacci sequences.

$$F_0=0,\ F_1=1,\ \text{when } n>1,\ F_n=F_{n-1}+F_{n-2}.$$

(4) Write a program that reads in an integer, which is greater than 3, and determines if it is a prime number.

2.2.2 DO WHILE statement

➢ Key Points

The form of **DO WHILE** statement:

```
DO WHILE (condition)
  statements
END DO
```

where *statements* are a sequence of executable statements, and *condition* is a logical expression. It repeats the *statements* while the condition is. TRUE. .

➢ Examples

Example 2-2-5:

Calculate the sum of 1 to 100 and output the answer.

Decode:

The range of numbers for this question is from 1 to 100. Therefore, "the number is less than 100" can be set as the condition, while the sum is updated by adding the number and the number is updated by adding 1 in each loop.

Code:

```
1    program ex2_2_5
2    implicit none
3      integer n, sum
4      sum= 0
```

```
5       n= 1
6       do while (n< = 100)
7           sum= sum+ n
8           n= n+ 1
9       end do
10      print* ,"the sum is ",sum
11  end
```

Line 4 indicates that the same as the DO loop, the sum should be assigned an initial value of 0.

Line 5 indicates that the initial value of n is 1.

Line 6—9 are DO WHILE statements with condition $n \leqslant 100$.

Line 7 is to calculate the sum in each loop.

Line 8 is to update n by adding 1.

Results:

```
the sum is         5050
Press any key to continue . . .
```

Example 2-2-6:

Write a program to calculate the factorial of n, output factorial with formatted output statement, when the factorial is less than 100.

Decode:

The same as Example 2-2-1, the factorials of n is $n! = 1 \times 2 \times 3 \times 4 \times \cdots \times n$. The condition for DO WHILE statement can be set as "factorial less than 100". n should be assigned an initial value of 1 first.

Code:

```
1   program ex2_2_6
2   implicit none
3       integer:: n= 1, f= 1
4       do while (f< = 100)
5           write(* ,'("the factorial of ",i2," is ", i3)') n, f
6           n= n+ 1
7           f= f* n
8       end do
9   end
```

Line 3 is to assign the initial value 1 to n and f with the declaration statement.

Line 4—8 are DO WHILE statements with condition $f \leqslant 100$.

Line 6 is to update n in each loop.

Line 7 is to calculate the factorial f.

Results:

```
the factorial of  1 is   1
the factorial of  2 is   2
the factorial of  3 is   6
the factorial of  4 is  24
Press any key to continue . . .
```

Example 2-2-7:

Write a program that reads in an integer, which is greater than 3, and determines if it is a prime number.

Decode:

For a prime, the only divisors of this integer are 1 and itself. Divid this integer n by the divisors from 2 to $n-1$. If all the remainder is not 0, it is a prime. For this question, the divisor can be set as a condition factor. If the remainder of integer n divided by the divisor is not 0 and the divisor is smaller than n, then this divisor will be updated by adding 1 in each loop. Finally, check the divisor. If the divisor is n, this integer is a prime. If the divisor is smaller than n, this integer is not a prime.

Code:

```
1    program ex2_2_7
2    implicit none
3       integer n, i
4       read* ,n
5       i= 2
6       do while (mod(n,i)/= 0.and.(i< n))
7          i= i+ 1
8       end do
9       if (i= = n) then
10         print* ,n, " is a prime"
11      else
12         print* ,n," is not a prime"
13      end if
14   end
```

Line 5 assigns the initial value 2 to divisor i, while the divisor is from 2 to $n-1$.

Line 6—8 are DO WHILE statements, judging the result of the condition.

Line 7 indicates that if i is not the divisor of n, i will be updated until $i=n$. If i is the divisor of n, this DO WHILE is finished.

Line 9—13 are IF statements with two branches to check the value of i. If i is not the divisor of n, it means that the integer n has a divisor other than 1 and itself. So $i==n$ can be used as the IF condition.

Results:

Input 5, the output result on screen will be as follows.

```
5
        5 is a prime
Press any key to continue . . .
```

➢ Practices

(1) Find out all multiples of 7 from 1 to 1000, and calculate the sum of all multiples.

(2) Write a program to calculate the Greatest Common Divisor of two positive integers.

(3) Read the daily precipitation data one by one until the data is equal to the missing value, output the accumulate precipitation without the missing value.

(4) Output the minimum of the Fibonacci sequences greater than 3000.

2.2.3 Loop control statements

➢ Key Points

(1) The loop control statement—**EXIT**

```
DO loop- variable= start, end[, step]
    statements-1
    EXIT
    statements-2
END DO
```

Usually, the **EXIT** statement in DO loop is executed with logical **IF** statements, such as:

```
DO loop- variable= start, end[, step]
    statements-1
    if(logical-expression) EXIT
    statements-2
END DO
```

In each **DO** loop, *statements*-1 is executed first. If the *logical-expression* in **IF** statement is. **TRUE.** , the **EXIT** statement is executed, and the control leaves the current **DO** loop. In this case, sta*tements*-2 will never be executed. If the *logical-expression* is. **FALSE.** , *statements*-2 will be executed. For the nested **DO** loops, any **EXIT** statement in the inner **DO** loop will bring the control out of the inner **DO** loop including the current **EXIT** statement.

(2) The loop control statement—**CYCLE**

```
DO loop- variable= start, end[, step]
    statements-1
    CYCLE
    statements-2
END DO
```

Similar to the **EXIT** statement, the **CYCLE** statement in **DO** loops is usually executed with logical **IF** statements, such as:

```
DO loop-variable= start, end[, step]
    statements-1
    if(logical-expression) CYCLE
    statements-2
END DO
```

In each **DO** loop, *statements*-1 is executed. If the *logical-expression* in **IF** statement is. **TRUE.** , the **CYCLE** statement is executed, and the control finishes the current loop and starts the next loop. If the *logical-expression* is. **FALSE.** , *statements*-2 will be executed.

> **Examples**

Example 2-2-8:

Output 1 to 10, except for 9.

Decode:

The CYCLE statement can be used to skip the ninth loop.

Code:

```
1    program ex2_2_8
2    implicit none
3      integer i
4      do i= 1,10
5        if (i= = 9) cycle
6        print* ,i
7      end do
8    end
```

Line 4 — 7 are DO loop statements. The numbers from 1 to 10 are set as the loop variables.

Line 5 is a logical IF statement. If *i* is 9, it will execute the tenth loop.

Results:

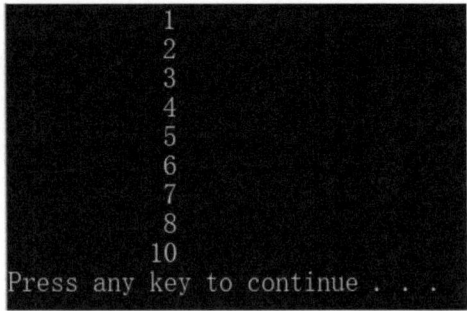

Example 2-2-9:

The same as Practice 4 of 2.2.2, output the minimum value of the Fibonacci sequences greater than 3000, with loop control statements.

Decode:

The Fibonacci sequences are $F_0 = 0$, $F_1 = 1$, when $n > 1$, $F_n = F_{n-1} + F_{n-2}$. The loop variable can be set much larger. In each loop, the Fibonacci sequences will be calculated. And, once the value of Fibonacci sequences is larger than 3000, exit the loop and output the value.

Code:

```
1    program ex2_2_9
2    implicit none
3      integer:: n,f1= 1,f2= 2,f
4      f= f1+ f2
5      do n= 1,100
6        if (f> = 3000)exit
7        f1= f2
8        f2= f
9        f= f1+ f2
10     end do
11     print* ,"the minimum is ",f
12   end
```

Line 3 is to declare and assign the initial value to the first two values of Fibonacci sequences $f1, f2$.

Line 4 is to calculate the third value of Fibonacci sequences.

Line 5—10 are the DO statements that extend the range of the loop variable to 100, in order to ensure f is larger than 3000.

Line 6 indicates that if the condition $f \geqslant 3000$ is. TRUE. , the control will leave the DO loop and execute the statement in Line 11.

Line 7—9 change the value of F_{n-1} and F_{n-2}, and update the f.

Results:

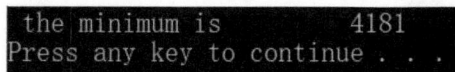

Example 2-2-10:

Read 10 days' temperature data, and calculate the average temperature. The missing value is 999.

DATE	01	02	03	04	05	06	07	08	09	10
Temp	23	26.5	999	24	22	21	29	25.5	999	27

Decode:

There are two missing values in the table. The missing value should be skipped in the calculation.

Code:
```
1    program ex2_2_10
2    implicit none
3      integer i, n
4      real temp, ave, total
5      total= 0.0
6      n= 0
7      do i= 1,10
8        read* ,temp
9        if(temp= = 999)cycle
10       total= total+ temp
11       n= n+ 1
12     end do
13     ave= total/n
14     print* ,"the average temperature is ",ave
15   end
```

Line 5—6 assign the initial value 0 to the number of days n whose temperature is not missing value and the sum of temperatures.

Line 7—12 are the DO loop with loop variables from 1 to 10.

Line 9 is the CYCLE statement with logical IF statement. Set the condition to whether *temp* is missing value, if the condition is. TURE. , the program will execute the next loop.

Line 10—11 update the *total* and *n*.

Results:

```
23
26.5
999
24
22
21
29
25.5
999
27
 the average temperature is    24.75000
Press any key to continue . . .
```

Example 2-2-11:

Write a program to output the prime from 100 to 150.

Decode:

Do the same as you did in Example 2-2-7, for a prime, the only divisors of this integer are 1 and itself. The nested DO statements can be used here. The outer loop variable is the integer from 100 to 150. The inner loop can be set to determine whether it is a prime. If the integer is not a prime, the outer loop will execute to the next loop.

Code:

```
1    program ex2_2_11
2    implicit none
3      integer i, j
4      print* ,"the primes between 100 to 150 are: "
5      do i= 100,150
6        do j= 2,i- 1
7          if (mod(i,j)= = 0)exit
8        end do
9        if (j= = i) print* ,i
10     end do
11   end
```

Line 5—10 are the outer loop with the number from 100 to 150 as the loop variables.

Line 6—8 are the inner loop with the divisor from 2 to $i-1$ as the loop variables.

Line 7 indicates that if the remainder of i and j is 0, i is not a prime. The program control will stop the inner loop and execute Line 9.

Line 9 indicates that if the remainder of i and j is not 0 for the whole inner loop, the j will be updated by adding 1 after the final inner loop, which means $j=i$. To judge whether j is equal to i with a logical IF statement, if the result is. TRUE. , i is a prime.

Results:

```
the primes between 100 to 150 are:
        101
        103
        107
        109
        113
        127
        131
        137
        139
        149
Press any key to continue . . .
```

➤ Practices

(1) Write a program to calculate and output the factorial of 3, 13, 23, ⋯,93.

(2) Input a real value x and compute $\exp(x)$ with the following formula until the absolute value of a term is less than 0.00001.

$$e^x = 1 + x + \frac{x^2}{2!} + \frac{x^3}{3!} + \cdots + \frac{x^n}{n!}$$

Chapter 3   Array

An array is an elements' collection of the same type with a fixed-size. Array elements are subscripted.

Terminology of array:
- *rank*: the number of dimensions that an array has.
- *extent*: the range of the subscripts of array elements.
- *bounds*: upper and lower limits of subscripts.
- *shape*: rank and extent of an array.
- *size*: the total number of elements or the number of elements in a given dimension.

The form of *extent* of an array is similar to the *list* in the **SELECT CASE** statement but only of **INTEGER** type:

```
smallest-subscript: largest-subscript
```

The *subscripts* of elements must be integers within the *extent*. The smallest and the largest subscripts are referred to as the lower and upper bound, respectively. If the lower bound of an *extent* is 1, it can be omitted as well as the colon following it.

3.1   One-dimensional array

> ➢ **Key Points**

(1) Declaration of one-dimensional array

Before using an array, the *name*, *type*, *dimension*, and *size* of the array must be specified so that the compiler can allocate the appropriate storage unit for the array. In FORTRAN 90/95, there are mainly three methods for defining arrays.

1) Type declaration statement

```
type   array-name(s-sub: l-sub)
```

where *type* can be **INTEGER**, **REAL**, **COMPLEX**, **LOGICAL** and **CHARACTER**, *s-sub* and *l-sub* are the *smallest-subscript* and *largest-subscript*, which show the range of subscripts of elements with **INTEGER** type.

For example:

```
integer a(-5:5), b(20)
character* 5  c(20)
```

The above statements define two one-dimensional integer arrays a, b and a one-dimensional character array c. The *ranks* of arrays a, b and c are 1. The *extent* of a is 11, while b and c are 20. The *subscript* of the array a ranges from -5 to 5 with *lower bound* -5 and *upper bound* 5. Array b and array c range from 1 to 20 with lower bound 1 and upper bound 20. The *size*, as total elements, of array a is 11, while b and c are 20. The *shape* of array a is (/11/), b and c are (/20/). c is a character array, with 20 array elements, each of which can store a string of length 5.

2) **DIMENSION** statement

$$\text{DIMENSION \quad array-name(s-sub: l-sub)}$$
$$\text{type array-name}$$

DIMENSION statement specifies the *name* and *extent* of an array, then declares the type.

• The **DIMENSION** statement is a non-executable statement and must be placed before the executable statements in the program unit.

• If the array *type* is not specifically specified, the array *type* follows the I-N rule.

For example, the following array declarations:

```
dimension a(-5:5),b(20),c(20)
   integer a, b
   character* 5 c
```

The arrays defined here using **DIMENSION** are the same as the arrays defined with the above type declaration statement.

3) **DIMENSION** and type declaration statement

It generally has two ways of defining it:

```
type, DIMENSION(s-sub:l-sub):: array-name
type, DIMENSION:: array-name(s-sub:l-sub)
```

For example, the following array declarations:

```
integer, dimension(-5:5):: a
integer, dimension:: a(-5:5)
```

Above, one one-dimensional integer array a is defined using two different forms, with the array subscripts ranging from -5 to 5, with a total of 11 array elements.

```
integer, dimension(-3:3):: a(-5:5)
```

For the above method, if the subscript range followed by the array name is different from that followed by **DIMENSION**, then array a should be defined with the *extent* followed by the array name.

(2) Storage structure of the one-dimensional array

The storage structure of the one-dimensional array is the same as its logical structure, like a vector. It can be visualized as below:

```
       ┌──────────┐
       │ element-1│
       ├──────────┤
       │ element-2│
       └──────────┘
           ...
       ┌──────────┐
       │ element-n│
       └──────────┘
```

(3) Reference to array elements of the one-dimensional array

The elements have the same name, but can be distinguished by subscripts. The multiple elements also can be referenced at once.

1) Single array element reference

$$\text{array-name (subscript)}$$

where *array-name* is the name of the array, the *subscript* is an integer expression, and the value is between the *smallest-subscript* and the *largest-subscript*.

For example, here is an array defined as follows:

$$\text{integer a(5)}$$

Then the elements of array *a* are: $a(1)$, $a(2)$, $a(3)$, $a(4)$, $a(5)$. $a(3)$ represents the third element of the array.

2) Multiple array elements reference

$$\text{array-name (smaller-sub: larger-sub [: step])}$$

where *smaller-sub* represents the starting subscript, *larger-sub* represents the ending subscript, and *step* represents the interval. All three subscript expressions must be **INTEGER** expressions and can be omitted. If *smaller-sub* is omitted, the lower bound of the dimension is the *smallest-subscript* of the array; if *larger-sub* is omitted, the upper bound of the dimension is the *largest-subscript* of the array; if *step* is omitted, *step* is 1.

• The *smaller-sub* and *larger-sub* must be between the *smallest-subscript* and the *largest-subscript*.

• *Step* must be less than *larger-sub* minus *smaller-sub*.

For example, here is an array defined as follows:

$$\text{integer a(5)}$$

$a(2:4:1)$ represents the three array elements $a(2)$, $a(3)$, $a(4)$

$a(:5:2)$ represents the three array elements $a(1)$, $a(3)$, $a(5)$

(4) Input and output of the one-dimensional array

1) Assignment with DATA statement

$$\text{DATA array-name/values/}$$

where **DATA** is a keyword, *array-name* is the name of the array, and *values* between two '/' are the initial values for each element.

For example:

$$\text{integer a(3)}$$
$$\text{data a/5,3,1/}$$

The above program defines a one-dimensional array a, and assigns initial values to all array elements in array a. That is, $a(1)=5$, $a(2)=3$, $a(3)=1$.

The initial value table can use constants, symbolic constants, constant expressions, and implicit **DO** loops, but not variables.

2) Assignment with array assignment operator

```
type:: array-name(s-sub:l-sub)= (/values/)
```

The initial values can be assigned in the declaration part.

For example:

```
integer:: a(3)= (/5,3,1/)
```

The initial values assigned to the one-dimensional array above are the same as those assigned with **DATA** statement.

3) Input and output the array elements with **DO** statement or implied **DO** statement

The implied **DO** statement form:

```
(list, loop- var= start, end[, step])
```

where *list* is the list of variables, expressions, and constants. *loop-var = start, end[, step]* are the same as DO loop.

The two nested implied **DO** statements are:

```
((list,loop-var= start,end[,step]),loop-var2= start2,end2[,step2])
```

The outer and inner **DO** loops have different loop variables.

For example:

```
integer a(5),i
do i= 1,5,1
   read(* ,* ) a(i)
end do
print* ,(a(i),i= 1,5)
```

The above program implements the input of one-dimensional array a from the keyboard with the **DO** statement and the output to the screen through the implicit **DO** statement.

4) Input and output of all array elements with the array name

For example:

```
integer a(5)
read* , a
print* , a
```

Above, all the elements of array a are inputted simultaneously and outputted.

3.2 Multi-dimensional array

➢ Key Points

(1) Declaration of multi-dimensional array

Similar as the one-dimensional array, multi-dimensional arrays can be defined with three forms.

    1) Type declaration statement

```
type  array-name(s-sub-1:l-sub-1, ⋯, s-sub-n:l-sub-n)
```

where *type* can be **INTEGER**, **REAL**, **COMPLEX**, **LOGICAL** and **CHARACTER**, *s-sub-i* and *l-sub-i* are the *smallest-subscript* and *largest-subscript* for the *ith* dimension, which show the *extent* of each dimension with **INTEGER** type. Different *extents* are separated by commas.

    For example:

```
integer a(2,3)
```

The array *a* is declared as a two-dimensional integer array. The *rank* of array *a* is 2. The *extents* of *a* are 2 and 3. The *subscripts* of *a* is ranging from 1 to 2 with *lower bound* 1 and *upper bound* 2 for the first dimension and 1 to 3 with *lower bound* 1 and *upper bound* 3 for the second dimension. The *size* of array *a* is 6, which means there are 6 elements in array *a*.

    2) **DIMENSION** statement

```
DIMENSION  array-name(s-sub-1:l-sub-1,⋯,s-sub-n:l-sub-n)
     type array-name
```

    The same as the one-dimensional array. Array *name* and *extent* are defined by **DIMENSION** statement. The *type* is set by the second statement.

For example, the following array declarations:

```
dimension a(2,3)
    integer a
```

Array *a* has the same characteristic as the example for the first declaration method.

    3) **DIMENSION** and type declaration statements

```
type, DIMENSION(s-sub-1:l-sub-1,⋯,s-sub-n:l-sub-n) :: array-name
type, DIMENSION:: array-name(s-sub-1:l-sub-1,⋯,s-sub-n:l-sub-n)
```

According to their characteristics, if you want to define multiple arrays of the same type and size, the first form is more convenient; if multiple arrays are of the same type but different sizes, the second method is more suitable.

    For example,

```
integer, dimension(2,3)::a
integer, dimension:: a(2,3)
```

The above statements declare the same two-dimensional array *a*.

    (2) Storage structure of multi-dimensional array

    The logical structure of a multi-dimensional array is a table (two-dimensional array) or a group of tables. Taking a two-dimensional array as an example, the logical structure is:

	Column1		Column2
Row1	element-1-1	...	element-1-n
Row2	element-2-1	...	element-2-n
...	...	...	...
Rown	element-n-1	...	element-n-n

the first dimension is the row, the second dimension is the column.

But the storage structure of multi-dimensional array is inconsistent with the logical structure, it is also shown as a vector, which stores along the column, is different from one-dimensional array:

element-1-1
element-2-1
...
element-n-1
element-1-2
element-2-2
...
element-n-2
...
...
element-1-n
element-2-n
...
element-n-n

(3) Reference to array elements

1) Single array element reference

    `array-name(subscripts-1,..., subscripts-n)`

where *array-name* is the name of the array, and the *subscripts-i* is an integer expression, the value is between the *smallest-subscript* and the *largest-subscript* for each dimension.

For example:

    `integer a(2,3)`

Array *a* is a two-dimensional array of integer type. There are 6 elements in array a, they are: $a(1,1)$, $a(1,2)$, $a(1,3)$, $a(2,1)$, $a(2,2)$, $a(2,3)$. If the element is in the second row and second column, we can represent it as $a(2,2)$.

2) Multiple array elements reference

```
array-name(smaller-sub-1:larger-sub-1:step-1,…,
          smaller-sub-n:larger-sub-n:step-n)
```

where *smaller-sub-i* represents the starting subscript, *larger-sub-i* represents the ending subscript, and *step-i* represents the interval for the *ith* dimension.

For example:

```
integer a(2,3)
a(1:2:1,1:3:2) represents a(1,1),a(1,3),a(2,1),a(2,3)
a(::,::2) represents a(1,1), a(1,3), a(2,1),a(2,3)
```

(4) Input and output of the multi-dimensional array

1) Assignment with **DATA** statement

```
DATA array-name/values/
```

where **DATA** is a keyword, *array-name* is the name of the array, *values* are the initial values for each element with **storage structure**.

For example:

```
integer a(2,3)
data   a/2,4,6,5,3,1/
```

The program above defines a two-dimensional array a, and assigns initial values to all array elements in array a. That is, $a(1,1)=2$, $a(2,1)=4$, $a(3,1)=6$, $a(1,2)=5$, $a(2,2)=3$, $a(3,2)=1$.

2) Assignment with array assignment operator

```
type:: array-name(s-sub-1:l-sub-1,…,s-sub-2:l-sub-2)= (/values/)
```

For example:

```
integer:: a(2,3)= (/ 2,4,6,5,3,1/)
```

The initial values assigned to the two-dimensional array above are the same as those assigned with **DATA** statement.

3) Input and output array elements with **DO** statement or implied **DO** statement

For example:

```
integer i,j
real a(5,3)
do i= 1,5,1
   read(* ,* ) (a(i,j),j= 1,3)
end do
print* , ((a(i,j),j= 1,3),i= 1,5)
```

The program above implements the input of two-dimensional array a with the keyboard and the output to the screen through the **DO** statement and the implicit **DO** statement.

4) Input and output of all array elements with the array name
For example:
$$\text{real a(5,3)}$$
$$\text{read* , a}$$
$$\text{print* ,a}$$
Similar to the one-dimensional array, all elements of array a are input simultaneously.

➢ Examples
Example 3-1:

Define a one-dimensional integer array a, a one-dimensional string array b, and a two-dimensional real array c. The array a has 5 array elements, the first three array elements are the number 8, and the last two are inputted with the keyboard. The array b has 4 array elements, each array element can contain 8 characters, the first two elements are "HELLO", "FORTRAN", and the last two are inputted with the keyboard. The array c is a two-dimensional array with two rows and three columns. The first row is all 8, the second row is inputted from the keyboard. Then, print the three arrays to the screen separately.

Decode:

Firstly, the three arrays should be declared with different types and different sizes. Then, the given array elements can be assigned directly with assignment statements, and the last elements inputted with list-directed input statements. Finally, output all elements in the storage structure of each array with list-directed output statements.

Code:

```
1     program ex3_1
2     implicit none
3       integer i
4       integer a(5)
5       character* 8 b(4)
6       real c(2,3),M
7       a(1:3)= 8
8       b(1)= 'HELLO'
9       b(2)= 'FORTRAN'
10      c(1,:)= 8.0
11      do i= 1,2
12         read* , a(i+ 3),b(i+ 2)
13      end do
14      read* , M
15      c(2,:)= M
16      print* ,"Array a is: ", a
17      print* ,"Array b is: ", b
18      do i= 1,2
```

```
19        print* , "Row ",i, " of array c is: ", c(i,:)
20     end do
21     end
```

Line 3 declares i as a loop variable.

Line 4 declares a as an integer array with subscripts from 1 to 5. It has 5 array elements.

Line 5 declares b as a string array with subscripts from 1 to 4, 4 array elements, each array element's length is 8.

Line 6 declares c as a real two-dimensional array with two rows and three columns, with a total of 6 array elements, and declares M as a real variable.

Line 7 assigns the value of 8 to elements $a(1)$, $a(2)$, $a(3)$.

Line 8 assigns the "HELLO" to $b(1)$.

Line 9 assigns the "FORTRAN" to $b(2)$.

Line 10 assigns the value of 8 to the first row of the two-dimensional array c.

Line 11−13 use a DO loop to read the remaining elements of arrays a and b from the keyboard simultaneously.

Line 14−15 read the value of M from the keyboard, and then assign M to the second row of the array c.

Line 16−21 are to output the arrays a, b, and c to the screen respectively.

Results:

Input 4, CYJ, 2, MAKE, 5.0, the output result on screen will be as follows.

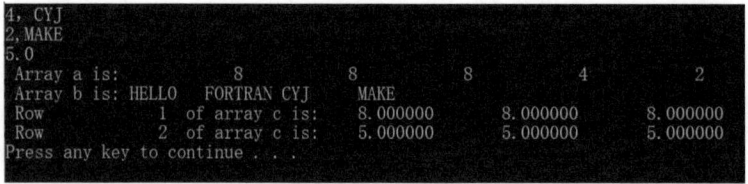

Example 3-2:

The table below shows the math test scores and students' IDs of 10 students. Please define the corresponding arrays based on the contents of the table, and assign the contents of the table to the arrays. Finally, print the IDs and scores of the students who scored above the average on the screen.

Student ID	01	02	03	04	05	06	07	08	09	10
Scores	59.5	65	75.5	100	80	45	96.5	74	85	32

Decode:

According to the problem, each student's data contains two items: student ID and math test score. The student ID can be represented by a string, and the score can be represented by a real number. Therefore, two one-dimensional arrays need to be defined, one is a string array to save the student IDs of 10 students, and the other is a real array to store the scores.

In addition, two real variables need to be defined to store the total and average scores. An IF statement is used to select the scores greater than average.

Code:

```
1    program ex3_2
2    implicit none
3      real::score(10)= (/59.5,65,75.5,100,80,45,96.5,74,85,32/)
4      character* 4 ID(10)
5      real sum, ave
6      data ID/'01','02','03','04','05','06','07','08','09','10'/
7      sum= 0.0
8      do i= 1,10
9         sum= sum + score(i)
10     end do
11     ave= sum/10
12     print* ,"The average score is:", ave
13     print* ,"Student ID scores"
14     print* ,"- - - - - - - - - - - - - - - - - - - - - - - - -"
15     do i= 1,10
16        if(score(i)> ave) print* ," ", ID(i), " ", score(i)
17     end do
18     print* ,"- - - - - - - - - - - - - - - - - - - - - - - - -"
19   end
```

Line 3 declares *score* as a one-dimensional real array with subscripts from 1 to 10, and assigns the scores in the table to this array.

Line 4 declares ID as a one-dimensional character array with an extent from 1 to 10, and each array element can store a string of length 4.

Line 5 declares *sum* and *ave* as real variables to save the total score and average.

Line 6 is to assign the values of Student ID in the table to the array ID with DATA statement.

Line 7 is to assign an initial value of 0.0 to *sum*.

Line 8—10 use a DO loop to accumulate the sum of the array *score*.

Line 11 is to calculate the average score.

Line 12 print the average score on the screen.

Line 15—17 output the scores and student IDs of students who scored above the average to the screen with a logical IF statement nested in a DO loop.

Results:

```
The average score is:    71.25000
Student ID   Scores
-----------------------------------
    03          75.50000
    04         100.0000
    05          80.00000
    07          96.50000
    08          74.00000
    09          85.00000
-----------------------------------
Press any key to continue . . .
```

Example 3-3:

Given a 3×3 matrix, find the maximum value of each column and its row number.

$$\begin{bmatrix} 5 & 7 & 10 \\ 9 & 6 & 4 \\ 1 & 0 & 2 \end{bmatrix}$$

Decode:

For a 3×3 matrix, a 3×3 two-dimensional integer array should be defined first, then assign the matrix values to the two-dimensional array. To find the maximum value, each column can be treated as a one-dimensional array and the row number of its maximum value can be found respectively.

Code:

```
1    program ex3_3
2    implicit none
3    real:: a(3,3)= (/5,9,1,7,6,0,10,4,2/), mamu
4    integer i, j, cmi
5    do j= 1,3
6       mamu= 0
7       cmi= 1
8       do i= 1,3
9          If (a(i,j)> mamu) then
10             mamu= a(i,j)
11             cmi= i
12         end if
13      end do
14      write(* ,'(1x,"The maximum value of column",i2,"is", I3,"it is
         in row",i2)') j, mamu, cmi
15   end do
16   end
```

Line 3 declares array *a* as a two-dimensional array and assigns the initial values in the storage structure, and *mamu* is a real variable to save the maximum.

Line 5—15 are the outer loop with *j* for column as the loop variable. In the outer loop, each column of array *a* is treated as a one-dimensional array.

Line 6—7 set the maximum value of each column to 0 and the maximum row number to 1.

Line 8—13 indicate the inner loop with *i* for row as the loop variable.

Line 9 — 12 are the nested IF statements. If the array element is larger than the maximum *mamu*, assign the value of the array element to *mamu*, and its row number *i* to *cmi*.

Line 14 outputs the maximum value and the position with given format.

Results:

```
The maximum value of column 1 is  9 it is in row 2
The maximum value of column 2 is  7 it is in row 1
The maximum value of column 3 is 10 it is in row 1
Press any key to continue . . .
```

Example 3-4:

Given a one-dimensional array with 10 array elements, enter a number from the keyboard at random, and then search the array elements in this one-dimensional array to see if this number exists.

Decode:

A DO loop nested with an IF statement can be used to implement a one-to-one comparison between the keyboard input value and the array elements. Since the size of the array is already known, the number of loops is known. Declare an integer variable as a flag. If the input number exists in the array, output the number, set the flag as 1. So, if the flag is not 1, that means the number doesn't exist in the array.

Code:

```
1    program ex3_4
2    implicit none
3       integer:: a(10)= (/5,65,93,16,74,36,20,15,43,21/)
4       integer i, fd, x
5       write(* , "(10i3)")(a(i),i= 1,10)
6       print* ,"Enter the number you want to search for:"
7       read* , x
8       fd= 0
9       do i= 1,10
10         if(a(i)= = x)then
11            write(* ,"(1x,i3,' is No.',i3)") x, i
12            fd= 1
13         endif
14      end do
15      if(fd= = 0)print* ,x," NO FIND!"
```

16   end

Line 4 sets x as the input number, fd as the flag, which will be updated to 1 by x equal to any array element.

Line 7 reads the number to be searched from the keyboard.

Line 8 sets the initial value 0 to fd.

Line 9—14 loop the number of array elements, and compare the input number x with the array elements one by one. If x is the same as one element, output it, and set $fd=1$.

Line 15 is to judge the value of fd. If not found, $fd=0$, output "NO FOUND!".

Results:

Input 18 to x, the output on the screen as follows.

```
 5 65 93 16 74 36 20 15 43 21
Enter the number you want to search for:
18
          18  NO FIND!
Press any key to continue . . .
```

Example 3-5:

Given a one-dimensional array with 10 elements, sort the array elements in descending order and output the array elements.

Decode:

Firstly, find the largest element among the 10 array elements, and then swap it with the first array element. Then, find the largest element from the remaining 9 elements and swap it with the second array element. Repeat this process 9 times, and the array elements will be sorted from the largest to the smallest.

Code:

```
1     program ex3_5
2     implicit none
3     integer:: a(10)= (/5,65,93,16,74,36,20,15,43,21/)
4     integer i, j, fd, x
5     print* ,"The array a before sorting is:"
6     write(* ,"(10i3)") (a(i),i= 1,10)
7     do i= 1,9
8         x= i
9         do j= i+ 1,10
10            if(a(j)> a(x))x= j
11        end do
12        if(i/= x) then
13            fd= a(i)
14            a(i)= a(x)
15            a(x)= fd
16        endif
```

```
17      end do
18      write(* ,'(/"The array a after sorting is:"/10i3)') (a(i),i= 1,10)
19   end
```

Line 4 indicates that fd is declared to store the maximum element, and x is used to save the subscript of the maximum element.

Line 7—17 is the outer loop, and the search is performed 9 times.

Line 8 indicates that in each outer loop, x gets the initial subscript as the loop variable i.

Line 9—11 is the inner loop to find the subscript of the largest array element.

Line 12—16 determine whether the maximum subscript is equal to the current outer loop variable i, if not, swap the largest array element with the ith array element.

Line 18 outputs the result with given format. "/" means line feed.

Results:

```
The array a before sorting is:
  5 65 93 16 74 36 20 15 43 21

The array a before sorting is:
 93 74 65 43 36 21 20 16 15  5
Press any key to continue . . .
```

Example 3-6:

Given a 4×3 matrix, calculate the transpose matrix of this matrix.

$$\begin{bmatrix} 8 & 2 & 1 \\ 5 & 7 & 10 \\ 9 & 6 & 4 \\ 1 & 0 & 2 \end{bmatrix} \rightarrow \begin{bmatrix} 8 & 5 & 9 & 1 \\ 2 & 7 & 6 & 0 \\ 1 & 10 & 4 & 2 \end{bmatrix}$$

Decode:

First, define a two-dimensional array a to store the 4×3 matrix, and a two-dimensional array b to store the 3×4 transpose matrix. When transposing, put the first column elements of array a into the first row of array b, and put the second column elements of array a into the second row of array b, i.e. swap the row and column subscripts of the array, like $a(i,j)=b(j,i)$.

Code:

```
1    program ex3_6
2    implicit none
3      integer:: a(4,3)= (/8,5,9,1,2,7,6,0,1,10,4,2/)
4      integer b(3,4), i, j
5      print* ,"The original matrix before transposition is:"
6      do i= 1,4
7        write(* , "(3i3)") (a(i,j),j= 1,3)
8      end do
```

```
 9          do i= 1,3
10            do j= 1,4
11              b(i,j)= a(j,i)
12            end do
13          end do
14          print* ,"The original matrix after transposition is:"
15          do i= 1,3
16            write(* , "(4i3)")(b(i,j), j= 1,4)
17          end do
18        end
```

Line 3 declares array a as a two-dimensional array, and set the initial values following the storage structure.

Line 6—8 output the original matrix with a DO statement and implied DO statement.

Line 9—13 transpose each row of array a into each column of array b to achieve the transpose of matrix a.

Line 15—17 output the transposed matrix.

Results:

```
The original matrix before transposition is:
 8  2  1
 5  7 10
 9  6  4
 1  0  2
The original matrix after transposition is:
 8  5  9  1
 2  7  6  0
 1 10  4  2
Press any key to continue . . .
```

Example 3-7:

Given a 4×3 matrix a and a 3×5 matrix b as follows, calculate the product $c = a \times b$ of the two matrices.

$$a = \begin{bmatrix} 8 & 2 & 1 \\ 5 & 7 & 10 \\ 9 & 6 & 4 \\ 1 & 0 & 2 \end{bmatrix} \quad b = \begin{bmatrix} 75 & 88 & 15 & 91 & 31 \\ 25 & 21 & 72 & 63 & 20 \\ 9 & 12 & 11 & 42 & 52 \end{bmatrix}$$

Decode:

When multiplying matrix a and matrix b, it is necessary to multiply each element of a row of matrix a with each corresponding element of a column of matrix b and then sum them up. Therefore, to calculate the product of two matrices, a triple nested loop is required, where the outer loop controls the rows of a, the middle loop controls the columns of b, and the inner loop calculates each element of c.

Code:

```
1    program ex3_7
2    implicit none
3      integer:: a(4,3)= (/8,5,9,1,2,7,6,0,1,10,4,2/)
4      integer:: b(3,5)= (/75,25,9,88,21,12,15,72,11,91,63,42,31,20,52/)
5      integer i, j, k, c(4,5)
6      print* ,"The matrix a is:"
7      do i= 1,4
8         write(* , "(3i3)") (a(i,j), j= 1,3)
9      end do
10     print* ,"The matrix b is:"
11     do i= 1,3
12        write(* , "(5i3)") (b(i,j), j= 1,5)
13     end do
14     do i= 1,4
15        do j= 1,5
16           c(i,j)= 0
17           do k= 1,3
18              c(i,j)= c(i,j)+ a(i,k)* b(k,j)
19           end do
20        end do
21     end do
22     print* ,"The a* b matrix c is:"
23     do i= 1,4
24        write(* , "(5i5)") (c(i,j), j= 1,5)
25     end do
26   end
```

Line 14—21 use a triple nested loop, where the outer loop controls the rows of matrix a, the middle loop controls the columns of matrix b, and the inner loop calculates each element of matrix c.

Line 16 initializes the elements of c with 0.0.

Line 18 update the c with adding the product of kth row of a and kth column of b.

Results:

```
The matrix a is:
  8  2  1
  5  7 10
  9  6  4
  1  0  2
The matrix b is:
 75 88 15 91 31
 25 21 72 63 20
  9 12 11 42 52
The a*b matrix c is:
  659  758  275  896  340
  640  707  689 1316  815
  861  966  611 1365  607
   93  112   37  175  135
Press any key to continue . . .
```

Example 3-8:

Insert a number into an ordered sequence with 10 numbers, (1, 3, 5, 7, 9, 11, 13, 15, 17, 19), the sequence remains the order after the insertion.

Decode:

Considering that the data will increase after the insertion, the array size should be increased from 10 to 11. Therefore, another array is set to save the new elements, whose size will be incremented by 1 after inserting a number.

Code:

```
1    program ex3_8
2    implicit none
3      integer i, j, k, a(10), b(11)
4      data (a(i),i=1,10)/1,3,5,7,9,11,13,15,17,19/
5      print* ,"The original matrix a is:"
6      write(* , "(10i3)") (a(i),i=1,10)
7      print* ,"Enter a number to insert:"
8      read* ,k
9      i=1
10     do while(k>a(i).and. i<=10)
11        i=i+1
12     end do
13     do j=1,10
14        if (j>i) b(j+1)=a(j)
15        if (j==i) b(j)=k
16        if (j<i) b(j)=a(j)
17     end do
18     print* ,"The inserted matrix b is:"
19     write(* , "(11i5)") (b(i), i=1,11)
```

20   end program

Line 10−12 compare the input data with 10 array elements of array a with the DO WHILE statement. Finally, determine its subscript in the new array with i.

Line 13−17 assign the input value to array b with subscript i. If the subscript is less than i, element subscripts of array b are the same as array a, otherwise, element subscript of array b are larger than array a by 1.

Results:

```
The original matrix a is:
 1  3  5  7  9 11 13 15 17 19
Enter a number to insert:
7
The inserted matrix b is:
     1    3    5    7    7    9   11   13   15   17   19
Press any key to continue . . .
```

➤ Practices

(1) One-dimensional array $a=(2.2, 1.5, 3.6, 4.9, 9.2)$, $b=$(I LOVE NUIST!), and the two-dimensional matrix c is as follows. Please define the above three arrays and output them on the screen.

$$c = \begin{bmatrix} 5 & 7 & 10 \\ 9 & 6 & 4 \\ 1 & 0 & 2 \end{bmatrix}$$

(2) Given the array declaration statement: integer $b(5,5)$, the contents of the array are as follows:

$$b = \begin{bmatrix} 1 & 2 & 4 & 6 & 8 \\ 3 & 0 & 8 & 7 & 2 \\ 5 & 8 & 5 & 9 & 3 \\ 7 & 1 & 2 & 3 & 0 \\ 9 & 6 & 1 & 2 & 5 \end{bmatrix}$$

Write FORTRAN program to implement the following functionalities:
• Input and output all array elements by row.
• Input and output all array elements by column.
• Output all array elements on the main diagonal.

(3) Input a one-dimensional array with 10 elements from the keyboard, then reverse it and output both the original and the reversed arrays. Rearrange the array elements in ascending order with the bubbling sorting method.

(4) The temperature values observed four times a day from July 17th to 21st at the Nanjing Meteorological Observatory are given in the following table:

Data/Time	00	06	12	18
17	28.9	31.9	36.9	33.3
18	29.9	32.4	36.1	31.2
19	28.8	33.5	35.2	32.4
20	30.5	32.6	36.2	32.8
21	30.0	33.0	36.6	25.6

Write FORTRAN program to implement the following functionalities:
• Calculate the average temperature for each day.
• Calculate the overall average temperature for the 5-day period.
• Output the maximum value of the average temperature for each day.
• Calculate the average temperature for each of the four observation times per day for the 5-day period.

(5) Calculate the standard deviation of the matrix elements in problem (2). The formula for standard deviation x_s is as follows:

$$x_s = \sqrt{\frac{\sum_{i=1}^{n}(x_i - x_a)^2}{n-1}}$$

where x_a is the average, n is the size of matrix, x_i is the *ith* element.

Chapter 4 Subprogram

Besides the main program, a complete program can also contain one or more subprograms. Fortran allows for two types of subprograms: **Functions** and **Subroutines**. In general, there are two forms of subprograms: **Internal** and **External**. Internal subprograms are those routines that may appear within the main program by making use of the **CONTAINS** statement. External subprograms are those which appear in a separate program unit after the main program **END** statement.

4.1 Function

In addition to intrinsic functions, Fortran allows users to design new functions. A Fortran function is a subprogram that implements a specific function, and returns the result, which is known as function value and returned by function name.

➢ **Key points**

(1) External function is an independent program unit

The form is:

```
FUNCTION function_name (dum_arg_list)
implicit none
    declaration statements
    executable statements
END [FUNCTION [function_name]]
```

• *function_name*: the name of the function.

• *dum_arg_list*: a comma-separated list of dummy arguments.

• *declaration statements*: one or more variable declarations, including those for the *function-name* and all arguments.

• *executable statements*: one or more statements that implement the function's functionality.

FUNCTION and **END** are the keywords that indicate the beginning and end of a function subprogram. Since the function returns its result via the *function_name*, the *function_name* should be declared in the *declaration statements*. The dummy arguments in *dum_arg_list* only have values from the actual arguments when the function is invoked. The function receives its input values from dummy arguments, does computations, and saves the result in its name. So, in the function, there must be one or more assignment statements as *function* _name=expression_ in the _executable statements_.

For the internal function, the function subprogram unit is placed before the **END** statement in the calling program unit. No function name declaration exists in calling program unit. There is the **CONTAINS** statement before the function unit as follows:

```
...
CONTAINS
FUNCTION function_name (dum_arg_list)
implicit none
    declaration statements
    executable statements
END FUNCTION [function_name]
...
```

(2) Function call

The way of using a function subprogram is the same as an intrinsic function, as:

```
function_name (act_arg_list)
```

The number, type, and positional order of the actual arguments in _act_arg_list_ must be consistent with the dummy arguments in the function subprogram unit. The names of the actual arguments and dummy arguments can be the same or not. The dummy argument only receives a value from its corresponding actual argument with its positional order.

➢ **Examples**

Example 4-1-1:

Do the same as you did in Example 1-3, write a program with a function to compute the geometric mean of three variables. The variables' values are inputted with the keyboard in the main program. Show the result on the screen. The geometric mean is $\sqrt[3]{a \times b \times c}$.

Decode:

The calculation statements are written in the function subprogram unit which is different from Example 1-3. The dummy arguments of the function are three variables. Their values are passed from the actual arguments in the function call statement of the main program.

Code:

```
1    program ex4_1_1
2    implicit none
3        real a, b, c, ave, gmean
4        read* , a, b, c
5        ave= gmean(a, b, c)
6        print* , ave
7    end


8    function gmean(x, y, z)
9        implicit none
```

```
10      real x, y, z, gmean
11      gmean= (x* y* z)* * (1.0/3.0)
12   end
```

Line 3 declares three real variables a, b, c, and the function name *gmean*. *ave* is used to save the result of the function call.

Line 4 inputs the variables' values with list-directed input statement.

Line 5 invokes the function with actual arguments a, b, c. The result returned by function name is assigned to *ave*.

Line 8 — 12 are the function subprogram unit. x, y, z are dummy arguments. The function name also should be declared. The value of actual argument a is passed to dummy argument x, the value of b is given to y, and z gets the value of c. Calculate the geometric mean of x, y, z and assign the result to function name *gmean*.

Results:

Input 1.2, 3.6, 9.0, and the output result on screen will be as follows.

```
1.2,3.6,9.0
   3.387730
Press any key to continue . . .
```

Example 4-1-2:

Please write a program in FORTRAN that includes an external function to compute the area of an ellipse. Input the length of the major and minor axes with the keyboard. The result is shown on the screen.

Decode:

The formula for calculating the area of an ellipse is π multiplied by the length of the minor axis and the length of the major axis. Therefore, the lengths of the minor and major axes of the ellipse can be used as the dummy arguments in the function unit. Their values are obtained from the actual arguments in the main program.

Code:

```
1    program ex4_1_2
2    implicit none
3       real a, b, area, ellipse_area_func
4       write(* ,* ) 'Enter the length of the major axis:'
5       read(* ,* ) a
6       write(* ,* ) 'Enter the length of the minor axis:'
7       read(* ,* ) b
8       area= ellipse_area_func(a, b)
9       write(* ,* ) 'The area of the ellipse is:', area
10   end

11   function ellipse_area_func(a, b)
```

```
12    implicit none
13      real a, b, pi, ellipse_area_func
14      pi= 4 * atan (1.0)
15      ellipse_area_func= pi * a * b
16    end
```

Line 3 declares *a* for the length of the major axis, *b* for the length of the minor axis with real type, and *area* for the area of the ellipse. The function name *ellipse_area_func* also be declared as a real variable. *a*, *b* are used as actual arguments for the function call.

Line 4—7 input the value to *a*, *b* with the keyboard.

Line 8 invokes the external function *ellipse_area_func* with actual arguments, and the value of actual arguments *a*, *b* will be passed to the dummy arguments to calculate the area of the ellipse.

Line 11 defines the external function *ellipse_area_func*, *a* and *b* are the dummy arguments of the function unit.

Line 13 declares the dummy argument, function name, and π.

Line 14—15 calculate the area of the ellipse, the result is returned by the function name.

Results:

```
Enter the length of the major axis:
9
Enter the length of the minor axis:
5
The area of the ellipse is:    141.3717
Press any key to continue . . .
```

Example 4-1-3:

Please write a program that includes an external function to compute the factorial of a number. Input the number with the keyboard. Output the result on the screen.

Decode:

The factorial of a number is the product of all positive integers from that number down to 1. Therefore, the external function only needs one input parameter.

Code:

```
1    program ex4_1_3
2    implicit none
3      integer n, result, i, factorial_func
4      write(* ,* ) 'Enter a positive integer:'
5      read(* ,* ) n
6      result= factorial_func(n)
7      write(* ,* ) 'The factorial of', n, 'is', result
8    end
9    function factorial_func(a)
```

```
10    implicit none
11      integer a, factorial_func, i
12      factorial_func= 1
13      do i= 1, a
14          factorial_func= factorial_func * i
15      end do
16    end function
```

Line 6 invokes the function, and n is the actual argument which gets the value from the keyboard before the function call.

Line 9 designs the function, and a is the dummy argument, and its value will be obtained from actual argument n.

Line 12—15 calculate the factorial with a DO loop.

Results:

```
Enter a positive integer:
8
The factorial of        8 is           40320
Press any key to continue . . .
```

Example 4-1-4:

Calculate the cross product of two one-dimensional vectors containing three elements with the internal function. Input and output are in the main program.

Decode:

The formula for vector cross product is:

$$(a_1, b_1, c_1) \times (a_2, b_2, c_2) = \begin{vmatrix} i & j & k \\ a_1 & b_1 & c_1 \\ a_2 & b_2 & c_2 \end{vmatrix} = (b_1 c_2 - b_2 c_1) i + (a_2 c_1 - a_1 c_2) j +$$

$$(a_1 b_2 - a_2 b_1) k = (b_1 c_2 - b_2 c_1, a_2 c_1 - a_1 c_2, a_1 b_2 - a_2 b_1)$$

The result is also a one-dimensional vector with 3 elements. The vector can be defined as a one-dimensional array.

Code:

```
1     program ex4_1_4
2     implicit none
3       integer  i
4       real   u(3), v(3), w(3)
5       print * , "Enter the components of vector u:"
6       read * , (u(i), i= 1,3)
7       print * , "Enter the components of vector v:"
8       read * , (v(i), i= 1,3)
9       w= cross(u, v)
10      print * , "The cross product of u and v is:"
```

```
11      print * , (w(i),i= 1,3)
12   contains
13      function cross(u, v)
14      implicit none
15         real, dimension(3):: u, v, cross
16         cross (1)= u(2)* v(3) - u(3)* v(2)
17         cross (2)= u(3)* v(1) - u(1)* v(3)
18         cross (3)= u(1)* v(2) - u(2)* v(1)
19      end function
20   end program
```

Line 9 invokes the function *cross*, and here u, v are actual arguments.

Line 12 is a CONTAINS statement for an internal function.

Line 13—19 are the internal function unit, and here u, v are dummy arguments.

Line 14 indicates that the function *cross* is also declared as an array.

Line 16—18 calculate the cross product, and the values of dummy arguments u, v are passed from the actual arguments u, v. Finally, return the result by function name in Line 9.

Results:

Input 2, 3, 4 to u and input 1, 0, 9 to v, and the output result on screen will be as follows.

```
Enter the components of vector u:
2
3
4
Enter the components of vector v:
1
0
9
The cross product of u and v is:
   27.00000        -14.00000        -3.000000
Press any key to continue . . .
```

➢ **Practices**

(1) Write a program that calculates the sum of two integers with an external function subprogram.

(2) The temperature values observed four times a day from July 17th to 21st at the Nanjing Meteorological Observatory are given in the following table:

Date/Time	00	06	18	20
17	28.9	31.9	36.9	33.3
18	29.9	32.4	36.1	31.2
19	28.8	33.5	35.2	32.4
20	30.5	32.6	36.2	32.8
21	30.0	33.0	36.6	25.6

Write an external function to achieve the following functions:
• Calculate the average temperature for each day.
• Calculate the overall average temperature for the 5-day period.
• Output the maximum value of each day's average temperature.

(3) Write an external function to determine if a number is prime. In the main program, input an integer and call the external function to output whether it is prime or not.

(4) Write an internal function to calculate the value of the following piecewise function:

$$y = \begin{cases} 3x^2 - \sqrt{1+x} & (x>0) \\ 2 & (x=0) \\ 3x + \sqrt{1-x} & (x<0) \end{cases}$$

4.2 Subroutine

A **Function** only returns one value with the function name, while **Subroutine** can also return no value or more than one value.

➤ **Key points**

(1) The external subroutine statement

```
SUBROUTINE
subroutine_name(dum_arg_list)
implicit none
    declaration statements
    executable statements
END [SUBROUTINE [subroutine_name]]
```

• *subroutine_name*: the name of the subroutine.

• *dum_arg_list*: a comma-separated list of dummy arguments.

• *declaration statements*: one or more variable declarations, including those for all arguments not *subroutine_name*.

• *executable statements*: one or more statements.

SUBROUTINE and **END** are the keywords that indicate the beginning and end of a subroutine subprogram unit. The same as the function, the dummy arguments only have values when it is invoked in the calling program unit. In a subroutine, there is no need to declare the *subroutine_name* in the subroutine and the calling program unit. The result can be returned with the arguments to the calling program unit.

For the internal subroutine, the subprogram unit is placed before the **END** statement in the calling program unit. There is the **CONTAINS** statement before the subroutine unit as follows:

```
...
CONTAINS
SUBROUTINE subroutine_name (dum_arg_list)
implicit none
    declaration statements
    executable statements
END SUBROUTINE [subroutine_name]
...
```

(2) The **CALL** statement

Unlike **Functions**, which can be used in expressions directly, **Subroutines** can only be called with the **CALL** statement.

```
CALL subroutine_name(act_arg_list)
```

The actual arguments in *act_arg_list* can be constants, variables, expressions, array names, array elements, or even subroutine names.

➤ **Examples**

Example 4-2-1:

Do the same as you did in Example 4-1-1, write a program with a subroutine to compute the geometric mean of three variables. The geometric mean is $\sqrt[3]{a \times b \times c}$.

Decode:

Different from Example 4-1-1, the calculation statements are written in the subroutine subprogram unit. The arguments are three variables and the result of the geometric mean calculation. There is no need to declare the name of the subroutine.

Code:

```
1   program ex4_2_1
2   implicit none
3     real a, b, c, ave
4     read* , a, b, c
5     call gmean(a, b, c, ave)
6     print* , ave
7   end

8   subroutine gmean(x, y, z, gave)
9   implicit none
10    real x, y, z, gave
11    gave= (x* y* z)* * (1.0/3.0)
12  end
```

Line 3 declares four real variables *a*, *b*, *c*, and *ave*. *ave* is used to save the result of the subroutine call.

Line 4 inputs the variables' value with a list-directed input statement.

Line 5 invokes the subroutine with actual arguments *a*, *b*, *c*, and *ave* by CALL statement. The calculation result is returned to *ave* from the corresponding dummy argument.

Line 8 — 12 are the subroutine subprogram unit. *x*, *y*, *z*, and *gave* are dummy arguments. The value of actual argument *a* is passed to dummy argument *x*, the value of *b* is given to *y*, and *z* gets the value of *c*. Calculate the geometric mean of *x*, *y*, *z*, and assign the result to *gave*. The value of *gave* is passed to the actual argument *ave*.

Results:

Input 1.2, 3.6, 9.0, the output result on screen will be as follows.

```
1.2, 3.6, 9.0
   3.387730
Press any key to continue . . .
```

Example 4-2-2:

Do the same as you did in Example 4-1-4, write the program to calculate the cross product of two one-dimensional vectors, but with the internal subroutine.

Decode:

The subroutine is different from the function, the results should be returned with the arguments. So, besides the arguments for input, one output argument should be used.

Code:

```
1    program ex_4_2_2
2    implicit none
3      real u(3), v(3), crossprod(3)
4      print * , "Enter the components of vector u:"
5      read * , u
6      print * , "Enter the components of vector v:"
7      read * , v
8      call cross(u, v, crossprod)
9      print * , "The cross product of u and v is: "
10     print * , crossprod
11   contains
12   subroutine cross(u, v, w)
13     implicit none
14       real, dimension(3) :: u, v, w
15       w(1) = u(2) * v(3) - u(3) * v(2)
16       w(2) = u(3) * v(1) - u(1) * v(3)
17       w(3) = u(1) * v(2) - u(2) * v(1)
18     end subroutine
19   end program
```

Line 3 indicates that besides *u* and *v*, *crossprod* is also declared as a real type array, which is the actual argument to save the calculation result of the subroutine.

Line 8 calls the subroutine to calculate the cross product of vectors with input actual arguments u, v, and output actual argument *crossprod*.

Line 12—18 are the internal subroutine with dummy arguments u, v, w. Here, w is the corresponding dummy argument of actual argument *crossprod*. The values of w will be passed to *crossprod*.

Results:

Input 2, 3, 4 to u and input 1, 0, 9 to v, the output result on screen will be as follows.

```
Enter the components of vector u:
2
3
4
Enter the components of vector v:
1
0
9
The cross product of u and v is:
   27.00000       -14.00000       -3.000000
Press any key to continue . . .
```

Example 4-2-3:

Write a program with a subroutine to calculate the value of a 10-degree polynomial with the given number x and polynomial coefficients.

$$p(x)=a_0+a_1x+a_2x^2+a_3x^3+\cdots+a_{10}x^{10}$$

Decode:

First, input the coefficients of the polynomial. The coefficients can be declared as an array with size n. Give a value for x. The x and coefficients can be used as input arguments. Another variable should be declared as an output argument to save the result of calculation. There are 11 coefficients for a 10-degree polynomial, so the coefficient array should be declared as a one-dimensional array with size 11.

Code:

```
1    program ex4_2_3
2    implicit none
3      integer:: i, n= 11
4      real x, y, coeff(11)
5      print * , "Enter the coefficients of the polynomial (from highest to lowest degree):"
6      read * , (coeff(i), i= n, 1, - 1)
7      print * , "Enter the value of x:"
8      read * , x
9      call poly(n, x, coeff, y)
10     print * , "The value of the polynomial at x is: ", y
```

```
11   end

12   subroutine poly(n, x, coeff, y)
13   implicit none
14     integer n, i
15     real x, coeff(n), y
16     y= coeff(n)
17     do i= n, 2, -1
18        y= y* x + coeff(i- 1)
19     end do
20     ! line 16- 20 also can be written as following
21     ! y= coeff(1)
22     ! do i= 2,11
23!        y= coeff(i)* x* * (i- 1)
24     ! end do
25   end subroutine
```

Line 4 indicates that because the number of coefficients is one more than the degree of the polynomial, here n is 11.

Line 6 reads in the coefficients of the polynomial from the highest to the lowest degree.

Line 8 reads in the value of x for the polynomial.

Line 9 calls the external subroutine *poly*.

Line 12 − 25 are a subroutine unit with a DO loop to calculate the values of each polynomial in two ways.

Results:

Input the coefficients 2, 3, 4, 5, 5, 6, 7, 8, 9, 10, 11 and 2 to x, the output result on screen will be as follows.

```
Enter the coefficients of the polynomial (from highest to lowest degree):
2, 3, 4, 5, 5, 6, 7, 8, 9, 10, 11
Enter the value of x:
2
The value of the polynomial at x is:      6003.000
Press any key to continue . . .
```

Example 4-2-4:

Write a program to transpose a 4×3 matrix.

Decode:

Transposing the column elements of array a into the row of array b, swap the row and column subscripts of the array, i. e. $a(i,j)=b(j,i)$.

Code:

```
1   program ex4_2_4
2   implicit none
3     integer, parameter:: n= 4, m= 3
```

```
4      integer i, j
5      real a(n, m), b(m, n)
6      print * , "Enter the elements of the matrix a:"
7      read * , ((a(i,j),j= 1,m),i= 1,n)
8      call transpose(n, m, a, b)
9      print * , "The transpose of the matrix a is:"
10     do j= 1, m
11        print * , (b(j,i), i= 1, n)
12     end do
13  end program

14  subroutine transpose(n, m, a, b)
15     implicit none
16       integer   n, m
17     real a(n,m)
18     real b(m,n)
19     integer i, j
20     do i= 1, m
21        do j= 1, n
22           b(i,j)= a(j,i)
23        end do
24     end do
25  end subroutine
```

Line10 reads in array a in the form of n rows and m columns.

Line 11 calls the external subroutine *transpose*.

Line 14—25 are a subroutine unit to implement matrix transposition.

Line 20—24 are nested DO loop statements to transpose the rows to columns.

Results:

```
Enter the elements of the matrix a:
8 2 1
5 7 10
9 6 4
1 0 2
 The transpose of the matrix a is:
    8.000000        5.000000        9.000000        1.000000
    2.000000        7.000000        6.000000        0.0000000E+00
    1.000000       10.00000         4.000000        2.000000
Press any key to continue . . .
```

➤ Practices

(1) Write a program to read the length and width in the main program and calculate the area of a rectangle, respectively in a subroutine.

(2) Develop a FORTRAN program that reads the following meteorological temperature data and calculates the average temperature for each day with internal subroutines.

Date/Time	00	06	12	18
17	28.9	31.9	36.9	33.3
18	29.9	32.4	36.1	31.2
19	28.8	33.5	35.2	32.4
20	30.5	32.6	36.2	32.8
21	30.0	33.0	36.6	25.6

(3) Write a FORTRAN program that calculates the dew point temperature from temperature and relative humidity data with an external subroutine. The value of temperature and relative humidity are read in the main program. The formula is as follows:

$$T_d = \ln\left(\frac{RH}{100}\right) + \left(\frac{17.27 \times T}{237.3 + T}\right) \bigg/ \left(\frac{17.27 - \ln\frac{RH}{100} + 17.27 \times T}{237.3 + T}\right)$$

Chapter 5 File

A file is a sequence of records. A record is a sequence of values or characters.

➢ **Key points**

(1) **OPEN** statement

The **OPEN** statement is used to open files for reading or writing.

$$\text{OPEN(list-of-specifiers)}$$

The primary *list-of-specifiers* contains the following specifiers:

Specifier	Value	Function	Default
UNIT	Integer: 0-99	File number	No default. When the file number is the first specifier, "unit =" can be omitted.
FILE	Character string	File name	No default. It can contain the complete directory.
ACCESS	Character string: SEQUENTIAL, DIRECT	File access mode	SEQUENTIAL
FORM	Character string: FORMATTED, UNFORMATTED, BINARY	Format type	Depends on ACCESS setting. FORMATTED for SEQUENTIAL files, but UNFORMATTED for DIRECT file.
RECL	Integer	Record length	No default.
STATUS	Character string: NEW, OLD, REPLACE, SCRATCH, UNKNOWN	File status	UNKNOWN
ACTION	Character string: READ, WRITE, READWRITE	File access	READWRITE

The simplest form of **OPEN** statements for different files:

- formatted sequential file

    OPEN (2, FILE= "A.DAT")

- formatted direct file

OPEN (2, FILE= "A.DAT", ACCESS= "DIRECT", FORM= "FORMATTED", RECL= 12)

- unformatted sequential file

OPEN (2, FILE= "A.DAT", FORM= "UNFORMATTED", ACCESS= "SEQUENTIAL")

- unformatted direct file

    OPEN (2, FILE= "A.DAT", ACCESS= "DIRECT", RECL= 16)

- sequential binary file

OPEN (2, FILE= "A.GRD", FORM= "BINARY")

(2) **CLOSE** statement

The **CLOSE** statement is used to terminate the connection between a logical unit and a file.

CLOSE (list-of-specifiers)

The *list-of-specifiers* has the following specifiers:

UNIT=*fnum*, file number.

IOSTAT=*ios* (optional).

STATUS=*sta* (optional). The status of the file after it is closed.

The simplest form: *CLOSE* (2), where 2 is a file number.

(3) **READ** & **WRITE** statements

The **READ** and **WRITE** statements are used for reading from and writing into a file respectively.

READ (list-of-specifiers) input table

WRITE (list-of-specifiers) output table

The *list-of-specifiers*: UNIT, FMT, REC, ADVANCE, SIZE, END, ERR, IOSTAT

The simplest form with UNIT (file number) and FMT (I/O format):

read(1,'(f3.2)') a

write(2,*) b

For BINARY file, only UNIT in *list-of-specifier*.

read(1) a

write(2) b

➤ **Examples**

Example 5-1:

Please write the station ID 23456, the date 20230520, and the temperature 26.3℃ to a sequential formatted file. Then, read them from the file.

Decode:

Firstly, assign the data to different variables and open a sequential formatted file. Secondly, write the data into this file with WRITE statements and close it. Thirdly, open this file again, read the data out with READ statements.

Code:

```
1    program ex5_1
2    implicit none
3      character* 8:: st= "23456"
4      character* 20:: date= "20230520"
5      real:: temp= 26.3
6      open (1,file= "st.txt")
7      write(1,"(a8)") st
8      write(1,"(a20)") date
```

```
9      write(1,"(f5.1)") temp
10     close(1)
11     open (2,file= "st.txt")
12     read(2,* ) st
13     read(2,* ) date
14     read(2,* ) temp
15     print * , st, date, temp
16     close(2)
17  end
```

Line 3—5 declare three variables to save stations' ID, date and temperature data and assign values.

Line 6 opens a sequential file first, and the file number is 1.

Line 7—9 write the data to file 1 with some given formats.

Line 10 closes the file 1.

Line 11 opens the file again but with the file number 2.

Line 12 — 15 read the data from file 2, and output them with list-directed output statement.

Results:

```
23456    20230520            26.30000
Press any key to continue . . .
```

Example 5-2:

Do the same as you did in Example 5-1, but write the data to a binary file. Then, read them out.

Decode:

Pay more attention to the form of OPEN and READ/WRITE statements for a binary file.

Code:

```
1   program ex5_2
2   implicit none
3     character* 8:: st= "23456"
4     character* 20:: date= "20230520"
5     real:: temp= 26.3
6     open (1, file= "st.dat", form= "binary")
7     write(1) st
8     write(1) date
9     write(1) temp
10    close(1)
11    open (2, file= "st.dat", form= "binary")
12    read(2) st
```

```
13      read(2) date
14      read(2) temp
15      print * , st, date, temp
16      close(2)
17    end
```

Line 6 opens a BINARY file first with "form='binary'" in the OPEN statement.
Line 7—9 write the data to file 1 only with file number as specifier.
Line 12—14 read the data from file 2 only with file number as specifier.

Results:

```
23456    20230520                26.30000
Press any key to continue . . .
```

> **Practices**

The same as Practice 7 in 4.1, the temperature values observed four times a day from May 17th to 21st, 2023 at the Nanjing are written in data.txt as follows:

Date/Time	00	06	12	18
17	28.9	31.9	36.9	33.3
18	29.9	32.4	36.1	31.2
19	28.8	33.5	35.2	32.4
20	30.5	32.6	36.2	32.8
21	30.0	33.0	36.6	25.6

Write a program to read the temperature data from the sequential formatted file, then calculate the temperature anomaly for each time and save them into a binary file "data.grd".

Chapter 6 One-dimensional Plot

Displaying two-dimensional data as a series of points, each point represents a pair of values. One on the x-axis and the other on the y-axis. These points can be connected by a line to show the overall feature of the data—***line plot***.

➤ **Key points**

(1) Line plot function

$$\text{gsn\_csm\_xy(wks, x, y, res)}$$

The arguments used in the function are set as follows:

• *wks*: A workstation identifier. The identifier is one returned either from calling **gsn_open_wks** or calling create to create a workstation object.

• *x/y*: The X and Y coordinates of the line(s). If x and/or y are two-dimensional, then the leftmost dimension determines the number of lines.

• *res*: A variable containing an optional list of plot resources, attached as attributes. Set to **True** if you want the attached attributes to be applied, and **False** if you either don't have any resources to set, or you don't want the resources applied.

(2) Common resource

1) xyLineColor / xyLineThicknesses / xyDashPattern

xyLineColor: This resource determines what color all the curves plot. You can use a color index value (integer) or a named color (string).

xyLineThicknesses: This resource sets the linewidth scale factor for curves.

xyDashPattern: This resource determines the dash pattern index of all the curves.

2) trYMinF / trYMaxF / trXMinF / trXMaxF

trYMinF / trYMaxF: Specify a minimum/maximum Y coordinate value that defines the lower/upper bound of the Y-Axis coordinate system.

trXMinF / trXMaxF: Specify a minimum/maximum X coordinate value that defines the upper bound of the X-Axis coordinate system.

3) tmXBMode / tmXTMode / tmYLMode / tmYLMode

tmXBMode / tmXTMode: This determines the method for specifying the spacing of the ticks and the contents of the labels along the bottom/top X-Axis. It has three possible settings: **"Automatic"**, **"Manual"** and **"Explicit"**.

tmYLMode / tmYLMode: As in *tmXBMode / tmXTMode*, but for left/right Y-axis.

4) gsnXRefLine / gsnYRefLine

Draw a horizontal line at the given X/Y value.

5) vpHeightF / vpWidthF

Specify the height/width of view.

6) tiMainString / tiXAxisString / tiYAxisString

Set the string to use as the Main title or X/Y-Axis title.

7) pmLegendDisplayMode / lgPerimOn

pmLegendDisplayMode: This resource determines whether the plot object displays a Legend object. It has four settings: **"NoCreate" "Never" "Always" "Conditional"**.

lgPerimOn: This resource determines whether to draw a line around the perimeter of the Legend.

> **Examples**

Example 6-1:

Draw the following line:

$$y = \sin(0.0628x)$$

x is from 0 to 100 and the step is 1.

Code:

```
1   begin
2   x= ispan(0,100,1)
3   y= sin(0.0628* x)
4   wks= gsn_open_wks("eps","ex6.1")
5   res= True
6   plot= gsn_csm_xy(wks,x,y,res)
7   end
```

Line 1 is the beginning of program.

Line 2—3 input x and y.

Line 4 opens a work station and sends graphics to file "ex6.1.eps".

Line 5 plots mods desired.

Line 6 creates a plot.

Line 7 is the end of program.

Results:

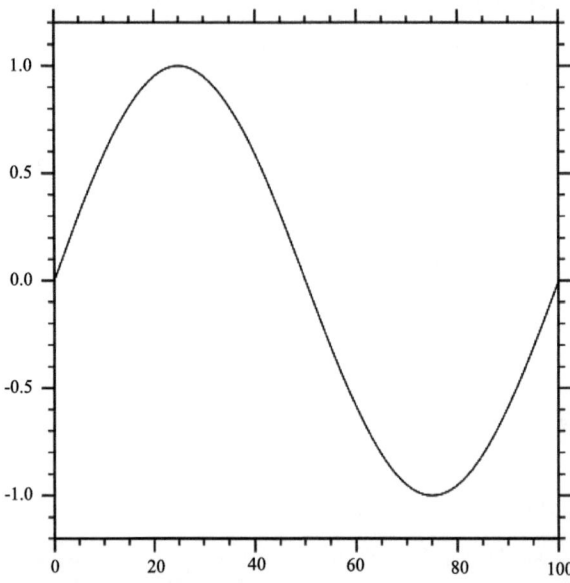

Example 6-2:

Change the color, thickness, and line type of the curve as given line in Example 6-1.

Code:

```
1   begin
2   x= ispan(0,100,1)
3   y= sin(0.0628* x)
4   wks= gsn_open_wks("eps","ex6.2")
5   res= True
6   res@ xyLineColor         = "Blue"
7   res@ xyLineThicknesses   = 2
8   res@ xyDashPattern       = 1
9   plot= gsn_csm_xy(wks,x,y,res)
10  end
```

Line 6 changes the line color to blue.

Line 7 sets the line thickness.

Line 8 sets the line type to dotted line.

Results:

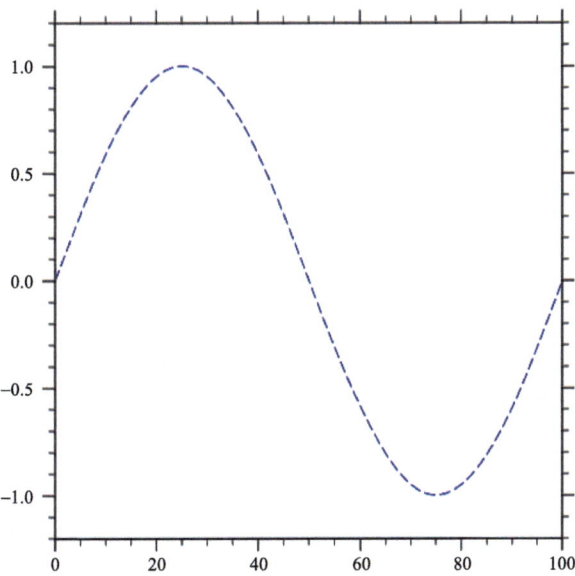

Example 6-3:

Adjust the range of the X or Y-axes in Example 6-1. While $x=0,25,50,75,100$. Set the corresponding labels as A, B, C, D, and E respectively.

Code:

```
1    begin
2    x= ispan(0,100,1)
3    y= sin(0.0628* x)
4    wks= gsn_open_wks("eps","ex6.3")
5    res= True
6    res@ trYMinF    = min(y)- 1
7    res@ trYMaxF    = max(y)+ 1
8    res@ trXMinF    = min(x)
9    res@ trXMaxF    = max(x)
10   res@ tmXBMode   = "Explicit"
11   res@ tmXBValues = (/0, 25, 50, 75, 100/)
12   res@ tmXBLabels = (/"A","B","C","D","E"/)
13   plot= gsn_csm_xy(wks,x,y,res)
14   end
```

Line 6—7 set minimum/maximum value of Y-axis as the minimum/maximum of y minus 1/plus 1.

Line 8—9 set minimum/maximum value of X-axis as the minimum/maximum of x.

Line 10 sets the mode of the bottom X-axis label.

Line 11—12 specify the bottom X-axis labels. When $x=0,25,50,75,100$, set the

corresponding labels as A, B, C, D, and E respectively.
   Results:

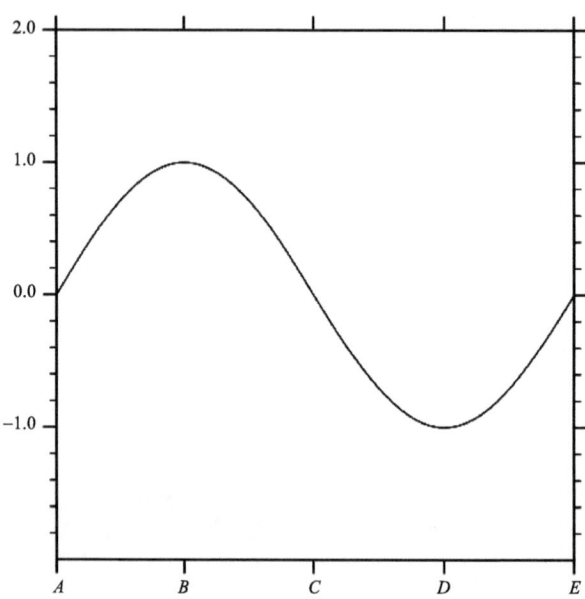

Example 6-4:

Add a horizontal line $y=0$ in Example 6-1. If $y>0$, color the part in red; if $y<0$, color the part in blue.
   Code:

```
1   begin
2     x= ispan(0,100,1)
3     y= sin(0.0628* x)
4     wks= gsn_open_wks("eps","ex6.4")
5     res= True
6     res@ gsnYRefLine              = 0
7     res@ gsnYRefLineColor         = "black"
8     res@ gsnYRefLineThicknesses   = 2
9     res@ gsnYRefLineDashPattern   = 0
10    res@ gsnAboveYRefLineColor    = "red"
11    res@ gsnBelowYRefLineColor    = "blue"
12    plot= gsn_csm_xy(wks,x,y,res)
13  end
```

Line 6 adds a horizontal line $y=0$.

Line 7 sets color of the reference line.

Line 8 sets thickness of the reference line.

Line 9 sets line type of the reference line.

Line 10 sets the parts above the reference line to fill in red.
Line 11 sets the parts below the reference line to fill in blue.
Results:

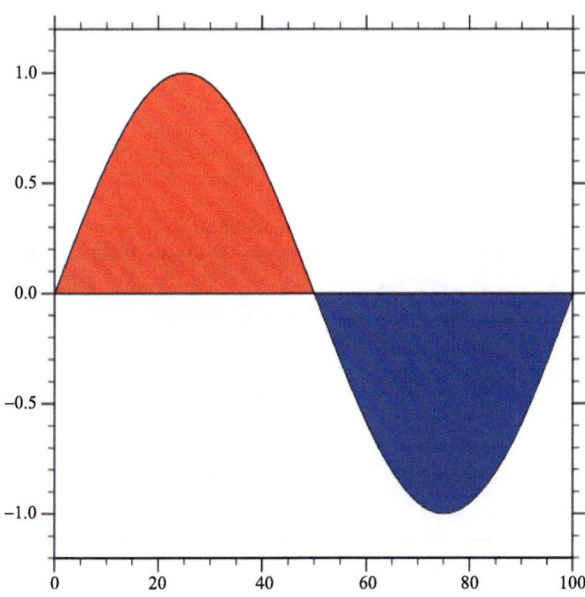

Example 6-5:

Adjust the image size to the appropriate ratio. Add a main title and X/Y-axis titles.

Code:

```
1    begin
2    x= ispan(0,100,1)
3    y= sin(0.0628* x)
4    wks= gsn_open_wks("eps","ex6.5")
5    res= True
6    res@ vpHeightF            = 0.3
7    res@ vpWidthF             = 0.6
8    res@ tiMainString         = "This is a main string"
9    res@ tiMainFont           = "helvetica-bold"
10   res@ tiMainFontHeightF    = 0.03
11   res@ tiXAxisString        = "X Values"
12   res@ tiXAxisFontHeightF   = 0.02
13   res@ tiYAxisString        = "Y Values"
14   res@ tiYAxisFontHeightF   = 0.02
15   plot= gsn_csm_xy(wks,x,y,res)
16   end
```

Line 6 sets the height of view.

Line 7 sets the width of view.

Line 8 adds a main title in image.
Line 9 sets font of the main title.
Line 10 sets font height of the main title.
Line 11 adds a title of X-axis.
Line 12 sets font height of the title of X-axis.
Line 13 adds a title of Y-axis.
Line 14 sets font height of the title of Y-axis.

Results:

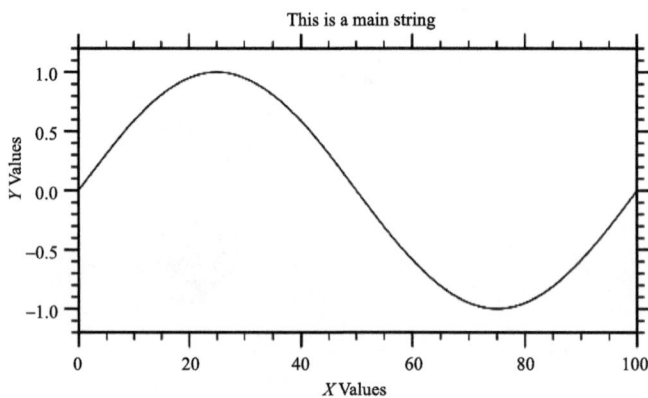

Example 6-6:

Draw the following lines in one image:

$$y1 = \sin(0.0628x)$$
$$y2 = \cos(0.0628x)$$
$$y3 = \sin(0.1628x)$$

x is from 0 to 100 and the step is 1. Create a legend and label each line with "y1", "y2" and "y3".

Code:

```
1   begin
2   x= ispan(0,100,1)
3   y1= sin(0.0628* x)
4   y2= cos(0.0628* x)
5   y3= sin(0.1628* x)
6   wks= gsn_open_wks("eps","ex6.6")
7   res= True
8   res@ xyLineThicknesses      = (/2,4,6/)
9   res@ xyDashPatterns         = (/0,4,8/)
10  res@ xyLineColors           = (/"blue","red","black"/)
11  res@ pmLegendDisplayMode    = "Always"
12  res@ pmLegendSide           = "Bottom"
```

13	res@ lgLabelFontThicknessF	= 2.0
14	res@ lgLabelFontHeightF	= 0.016
15	res@ lgPerimOn	= True
16	res@ pmLegendParallelPosF	= 0.15
17	res@ pmLegendOrthogonalPosF	= - 0.5
18	res@ pmLegendWidthF	= 0.08
19	res@ pmLegendHeightF	= 0.08
20	res@ xyExplicitLegendLabels	= (/" y1"," y2"," y3"/)
21	plot= gsn_csm_xy(wks,x,(/y1,y2,y3/),res)	
22	end	

Line 2—5 input x, $y1$, $y2$ and $y3$.

Line 8—10 set the specific thickness, color and line type of each line.

Line 11 turns legend on.

Line 12 sets location of the legend.

Line 13 sets font of the legend.

Line 14 sets font height of the legend.

Line 15 turns box around on.

Line 16 moves the units right, while positive means right.

Line 17 moves the units up, while positive means down.

Line 18 sets width of the legend.

Line 19 sets height of the legend.

Line 20 sets specific legend label.

Results:

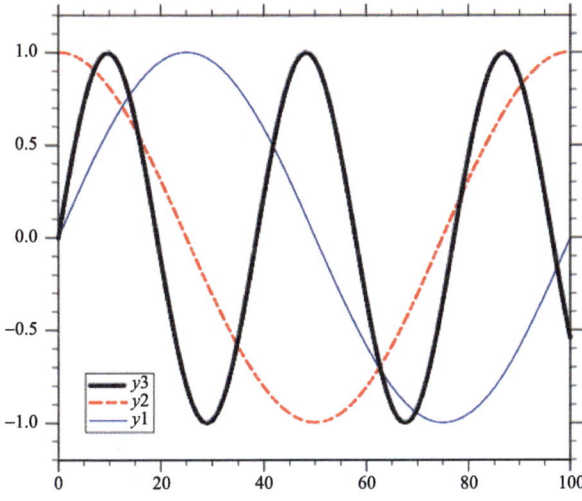

➤ **Practices**

(1) Draw the following line:

$$x=\cos(0.2784y^2)$$

y is from 0 to 100 and the step is 1. Add horizontal lines $x=0$, 1, -1. If $x>0$, color the part in red; if $x<0$, color the part in blue.

(2) Draw the following lines in one image:
$$x1=\cos(0.2784y^2)$$
$$x2=\cos(0.4539y^2)$$
$$x3=\sin(0.2784y^2)$$

y is from 0 to 100 and the step is 1. Create a legend and label $x1, x2, x3$ with "$x1$", "$x2$", and "$x3$", respectively. Besides, each line must be set to different colors, thicknesses, and line types.

Chapter 7 Two-dimensional Plot

7.1 Contours and shading

In NCL, a shading plot refers to a type of plot that is used to represent data on a two-dimensional grid, where each grid point is assigned a color based on its value. The shading plot is also known as a filled contour plot or a color-filled contour plot. —*shading plot*.

➢ **Key points**

(1) Create and draw a contour plot

$$\text{gsn\_csm\_xy(wks, x, y, res)}$$

The arguments used in function are set as follows:

• wks: A Workstation identifier. The identifier is one returned either from calling **gsn_open_wks** or calling create to create a workstation object.

• x/y: The data to contour; must be one-or two-dimensional.

• res: A variable containing an optional list of plot resources, attached as attributes. Set to **True** if you want the attached attributes to be applied, and **False** if you either don't have any resources to set, or you don't want the resources applied.

(2) Turn on color fill for a contour plot

$$\text{cnFillOn= True}$$

(3) Create and draw a contour plot over a map

$$\text{gsn\_csm\_contour\_map(wks,data,res)}$$

(4) Don't show contour lines

$$\text{cnLinesOn= False}$$

(5) Zoom in on map

$$\text{mpMinLatF}\backslash\ \text{mpMaxLatF}\backslash\ \text{mpMinLonF}\backslash\ \text{mpMaxLonF}$$

(6) Add string

$$\text{gsnLeftString}\backslash\ \text{gsnRightString}\backslash\ \text{gsnCenterString}$$

(7) Define a color map for the given workstation

$$\text{gsn\_define\_colormap(wks, color\_map)}$$

• *color_map*: Can be a string array of named colors, an array of red/blue/green (RGB) triplets, or a predefined color map. You can find suitable color_map in https://www.ncl.ucar.edu/Document/Graphics/color_table_gallery.shtml.

(8) Select the contour intervals displayed in a plot

cnLevelSelectionMode= "AutomaticLevels"\
"ManualLevels"\ "ExplicitLevels"

• "AutomaticLevels": Auto (default).

• "ManualLevels": Starting at *cnMinLevelValF*, contour levels are created at intervals spaced by the value of *cnLevelSpacingF* until *cnMaxLevelValF* is reached.

• "ExplicitLevels": This mode allows you to specify the value of each contour level by explicitly setting the contents of the *cnLevels* array.

➤ **Examples**

Example 7-1-1:

Draw the wind data with contour plot default settings. There are two wind components, u and v, with three dimensions (time, latitude, longitude). Plot the u component at the first time. The relevant wind data can be downloaded from this website: https://www.ncl.ucar.edu/Applications/Data.

Code:

```
1    begin
2    a= addfile("./uv300.nc","r")
3    u= a- > U(1,:,:)
4    wks= gsn_open_wks("pdf","ex7.1.1")
5    res= True
6    plot= gsn_csm_contour(wks,u,res)
7    end
```

Line 1 is the beginning of NCL file.

Line 2—3 read July zonal winds.

Line 4 sends graphics to PDF file.

Line 5 sets the desired plot mods.

Line 6 creates the plot.

Results:

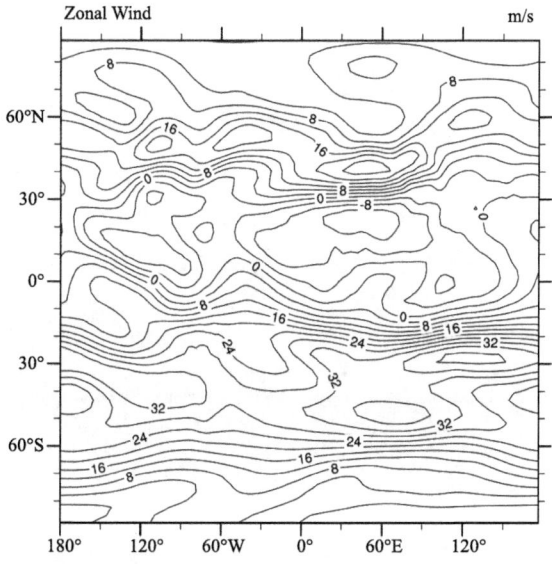

Example 7-1-2:

Do the same as you did in Example 7-1-1, but turn on the color with the default color map.

Code:

```
1    begin
2    a= addfile("./uv300.nc","r")
3    u= a- > U(1,:,:)
4    wks= gsn_open_wks("pdf","ex7.1.2")
5    res= True
6    res@ cnFillOn= True
7    plot= gsn_csm_contour(wks,u,res)
8    end
```

Line 6 turns on the color fill with the default colorbar.

Results:

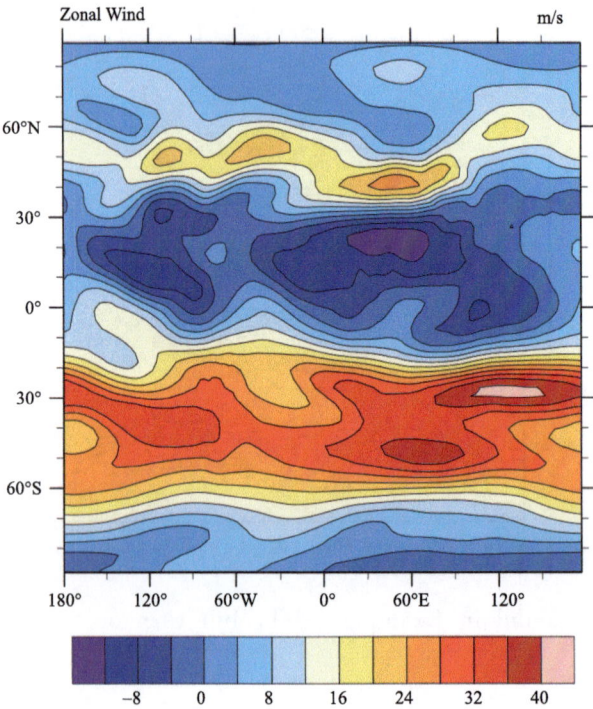

Example 7-1-3:

Do the same as you did in Example 7-1-2, but plot over a map.

Code:

```
1    begin
2    a= addfile("./uv300.nc","r")
3    u= a-> U(1,:,:)
4    wks= gsn_open_wks("pdf","ex7.1.3")
5    res= True
6    res@ cnFillOn  =  True
7    ;Don't show contour lines
8    res@ cnLinesOn= False
     ; creat a contour plot over a map
9    plot= gsn_csm_contour_map(wks,u,res)
10   end
```

Line 7 turns on color fill.

Line 8 sets no contour lines in the contour plot.

Line 9 creates a contour plot over a map.

Results:

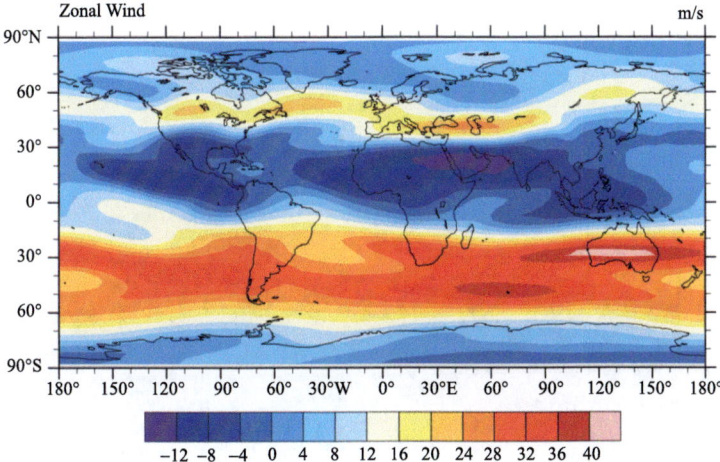

Example 7-1-4:

Do the same as you did in Example 7-1-3, but adjust the drawing range and add strings.

Code:

```
1    begin
2    a= addfile("./uv300.nc","r")
3    u= a- > U(1,:,:)
4    wks= gsn_open_wks("pdf","ex7.1.4")
5    res= True
6    res@ cnFillOn    = True
     ; zoom in on map
7    res@ mpMinLatF              = 0.0
8    res@ mpMaxLatF              = 60.0
9    res@ mpMinLonF              = 30.0
10   res@ mpMaxLonF              = 150.0
         ; Add strings
11   res@ gsnLeftString          = "U300"
12   res@ gsnRightString         = "m/s"
13   res@ gsnCenterString        = "Shade"
14   res@ gsnStringFontHeightF= 0.02
15   plot= gsn_csm_contour_map(wks,u,res)
16   end
```

Line 7 specifies the minimum latitude boundary.

Line 8 specifies the maximum latitude boundary.

Line 9 specifies the minimum longitude boundary.

Line 10 specifies the maximum longitude boundary.

Line 11 adds the given string just above the plot's upper boundary and left-justifies it.

Line 12 adds the given string just above the plot's upper boundary and right-justifies it.
Line 13 adds the given string just above the plot's upper boundary and centers it.
Line 14 resizes the string.

Results:

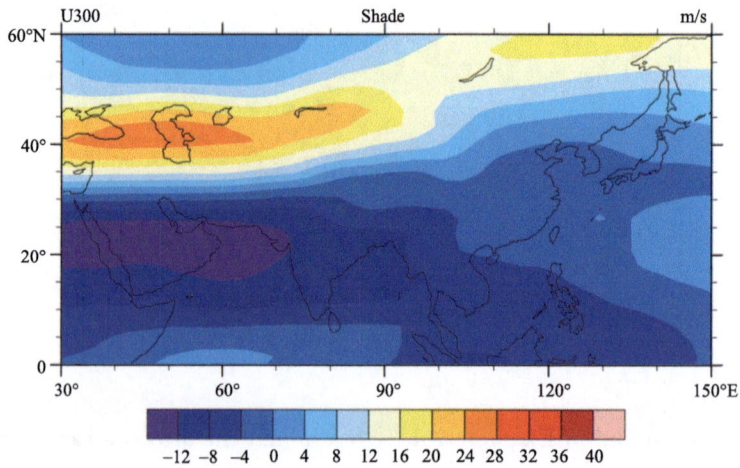

Example 7-1-5:

Do the same as you did in Example 7-1-3, but define a color map and select the contour intervals.

Code(way 1):

```
1    begin
2    a= addfile("./uv300.nc","r")
3    u= a- > U(1,:,:)           ; read July zonal winds
4    wks= gsn_open_wks("pdf","ex7.1.5.1")
     ;Define a color map for the given workstation
5    gsn_define_colormap(wks, "BlueDarkOrange18")
6    res= True
7    res@ cnFillOn     = True
8    res@ cnLinesOn    = False
     ; Select the contour intervals
9    res@ cnLevelSelectionMode= "ManualLevels"
10   res@ cnMinLevelValF     = -40
11   res@ cnMaxLevelValF     = 40
12   res@ cnLevelSpacingF    = 10
13   plot= gsn_csm_contour_map(wks,u,res)
14   end
```

Line 5 defines a color map for the given workstation.
Line 9 select "ManualLevels" as the contour intervals(Way 1).
Line 10 determines the lowest contour level.

Line 11 determines the highest contour level.
Line 12 determines the spacing between contour intervals.
Results:

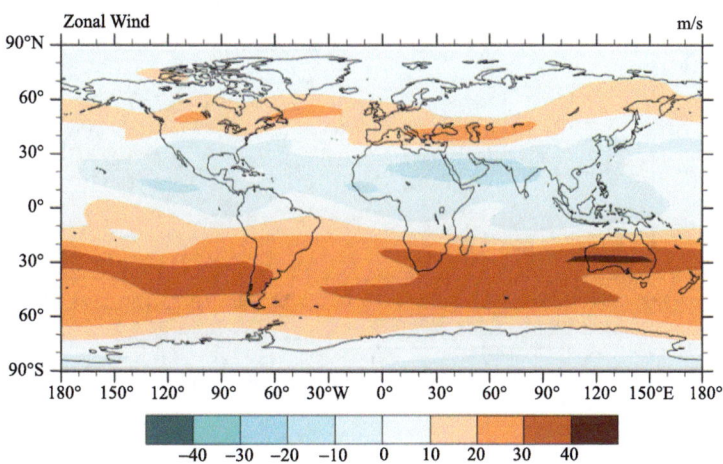

Code(way 2):

```
1    begin
2    a= addfile("./uv300.nc","r")
3    u= a- > U(1,:,:)
4    wks= gsn_open_wks("pdf","ex7.1.5.2")   ;Defines a color map for the
     given workstation
5    gsn_define_colormap(wks, "BlueDarkOrange18")
6    res= True
7    res@ cnFillOn   = True
8    res@ cnLinesOn  = False
     ; Selecting the contour intervals
9    res@ cnLevelSelectionMode = "ExplicitLevels"
10   res@ cnLevels              = (/-50,-40,-25,-10, 5,16, 48,62/)
11   res@ cnFillColors          = ispan(2,18,2)
12   plot= gsn_csm_contour_map(wks,u,res)
13   end
```

Line 5 defines a color map for the given workstation.

Line 9 select "ExplicitLevels" as the contour intervals (Way 2).

Line 10 indicates that an array of floats containing the contour levels is used to render the contour plot.

Line 11 represents an array of color indexes.

Results:

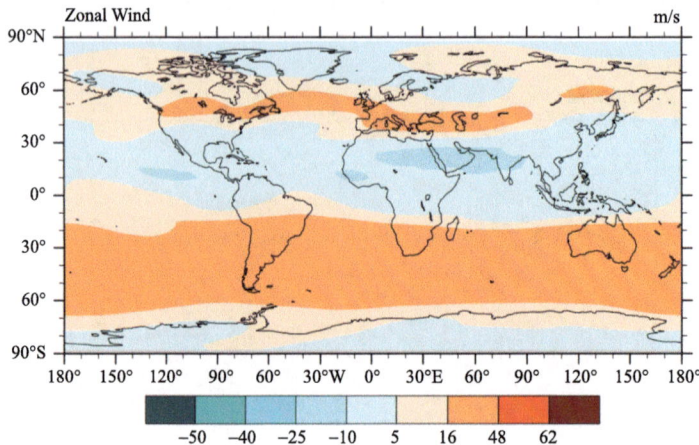

➢ **Practices**

Draw shadings for the following questions.

(1) Global July meridional wind at 300 hPa with the color table of BlueDarkRed18, without contour lines, by "ManualLevels".

(2) July meridional wind at 300 hPa over Europe, with contour lines over the shading, by "ExplicitLevels".

7.2 Vector figure

In NCL, you can use the "Vector" function to define a vector graph. You can use arrows to represent the direction and size of data. These vectors can simply represent speed, wind direction, or flow direction, and they can also be used for higher-level applications, such as tracking air mass movements or estimating turbulent transport, by creating vector maps that can visually present large and complex data sets and their changing trends, and can locate the outliers and then make effective judgment processing—***Vector plot***.

➢ **Key points**

(1) Create and draw a vector plot

```
gsn_csm_vector(wks,u,v, res)
```

The arguments used in the function are set as follows:

• *wks*: A workstation identifier. The identifier is one returned either from calling **gsn_open_wks** or calling create to create a workstation object.

• *u/v*: The *u* and *v* data for the vector plot; must be two-dimensional.

• *res*: A variable containing an optional list of plot resources, attached as attributes. Set to **True** if you want the attached attributes to be applied, and **False** if you either don't have any resources to set, or you don't want the resources applied.

(2) Create and draw vector maps on a map

gsn_csm_vector_map(wks,u,v,res)

(3) Vector field drawing parameters

vcGlyphStyle: This resource selects the style of glyph used to represent the vector magnitude and direction. There are four choices: *LineArrow* (default), *FillArrow*, *WindBarb*, *CurlyVector*.

vcMinDistanceF: This resource specifies a minimum distance in NDC space that is to separate the data locations of neighboring vectors. It represents the minimum distance between arrows.

vcMinMagnitudeF: This resource specifies a minimum magnitude for elements of the vector field. Vectors with magnitudes less than this value will not be rendered in the vector field plot.

vcLineArrowColor: This resource can change the vector color.

vcLineArrowThicknessF: This resource sets the thickness of the line used to draw vector line arrows. This resource has an effect only when *vcGlyphStyle* is set to *LineArrow* or *CurlyVector*.

vcPositionMode: This resource specifies how the vector arrow is positioned in the vector field plot relative to the actual data location. There are three possibilities: *ArrowHead* (The head of the arrow is placed at the data location). *ArrowCenter* (The center of the arrow is placed at the data location). *ArrowTail* (The tail of the arrow is placed at the data location).

(4) Set the reference vector

vcRefAnnoOn: Default: **True**. If this boolean resource is set to **False**, *VectorPlot* will not draw the reference vector annotation.

VcRefMagnitudeF: This resource can define vector reference magnitude and adjust the size of the vectors.

VcRefLengthF: This resource can define the length of the reference vector.

VcRefAnnoPerimOn: Default: **True**. This is a resource that determines whether *VectorPlot* will draw an outline around the perimeter of the box surrounding the reference vector annotation. If set **False**, no outline will be drawn.

vcRefAnnoString1On: Default: **True**. If this resource is set to **False**, *VectorPlot* will not display a string above of the reference arrow.

vcRefAnnoString1: This resource can draw a string above the reference vector.

vcRefAnnoString2On: Default: **True**. If this resource is set to **False**, *VectorPlot* will not display a string under the reference arrow.

vcRefAnnoString2: This resource can draw a string under the reference vector.

VcRefAnnoBackgroundColor: This resource sets the background color used to fill the box surrounding the vector reference annotation. If you do not want the box to be filled at all, set *vcRefAnnoBackgroundColor* to Transparent(-1).

VcRefAnnoOrthogonalPosF: This resource can vertically move the reference vector.

vcRefAnnoParallelPosF: This resource can horizontally move the reference vector.
vcRefAnnoArrowLineColor: This resource can change the reference vector color.
vcRefAnnoFontHeightF: This resource can set the font size of the reference vector label.

(5) Create and draw a vector plot over a polar stereographic map

gsnPolar: This resource controls what polar hemisphere is shown on polar stereographic plots. Set to either "SH" or "NH".

gsn_csm_vector_map_polar(wks,u,v,res): This resource draws a vector plot on a polar stereographic map.

(6) Set fill arrows style

vcFillArrowWidthF / vcFillArrowHeadXF / vcFillArrowHeadYF / vcFillArrowHeadInteriorXF / vcFillArrowFillColor / vcFillArrowEdgeColor

(7) Set color vectors based on magnitude

read_colormap_file(filename: string): This resource reads an NCL system colormap file or a user-defined colormap.

vcLevelPalette: This resource allows you to set a color palette from which the values assigned to *vcLevelColors* are chosen.

vcMonoLineArrowColor: This resource can set vectors' color based on magnitude. This resource has an effect only when *vcGlyphStyle* is set to LineArrow or CurlyVector.

> **Examples**

Example 7-2-1:

Do the same as you did in Example 7-1-1, but with vector default settings.

Code:

```
1    begin
2    a= addfile("./uv300.nc","r")
3    u= a- > U(1,:,:)
4    v= a- > V(1,:,:)
5    wks= gsn_open_wks("pdf","ex7.2.1")
6    res= True
7    plot= gsn_csm_vector(wks,u(::5,::5),v(::5,::5),res)
8    end
```

Line 1 is the beginning of program.

Line 2—4 open the file and read in data.

Line 5 opens a work station and sends graphics to PDF file.

Line 6 sets the desired plot mods.

Line 7 creates the plot.

Line 8 is the end of program.

Results:

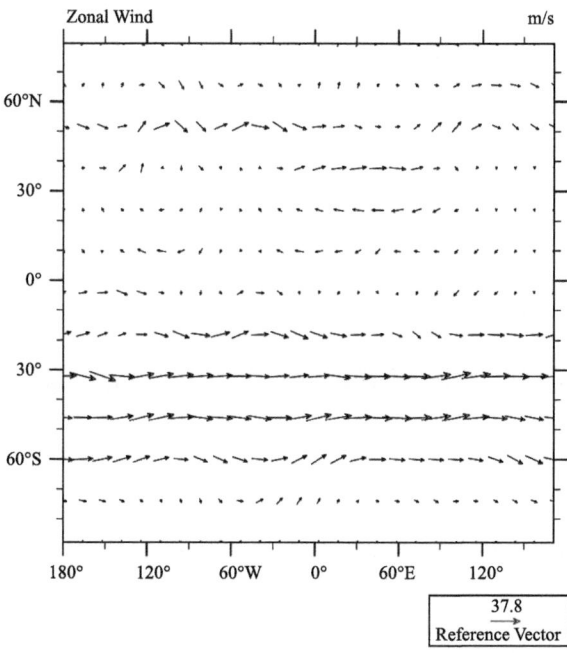

Example 7-2-2:

Do the same as you did in Example 7-2-1, but create a vector map on a map.

Code:

```
1   begin
2   a= addfile("./uv300.nc","r")
3   u= a- > U(1,:,:)
4   v= a- > V(1,:,:)
5   wks= gsn_open_wks("pdf","ex7.2.2")
6   res= True
7   plot= gsn_csm_vector_map(wks,u(::5,::5),v(::5,::5),res)
8   end
```

Line 7 creates a vector plot over a map.

Results:

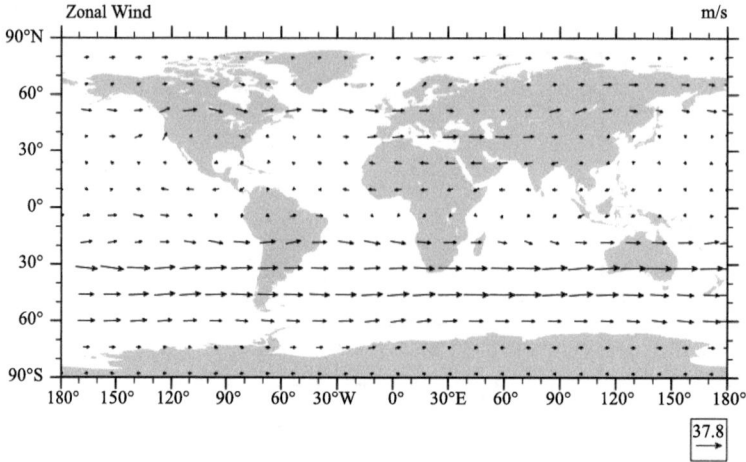

Example 7-2-3:

Do the same as you did in Example 7-2-2, set vector field drawing parameters.

Code:

```
1    begin
2    a= addfile("./uv300.nc","r")
3    u= a-> U(1,:,:)
4    v= a-> V(1,:,:)
5    wks= gsn_open_wks("pdf","ex7.2.3")
6    res= True
7    res@ gsnLeftString= ""
8    res@ gsnRightString= ""
9    res@ tmXTOn= False
10   res@ tmYROn= False
11   res@ vcVectorDrawOrder= "PostDraw"
12   res@ vcGlyphStyle= "CurlyVector"
13   res@ vcMinDistanceF= 0.02
14   res@ vcMinMagnitudeF= 0.5
15   res@ vcLineArrowColor= "black"
16   res@ vcLineArrowThicknessF= 2.5
17   res@ vcRefAnnoOn= True
18   res@ vcRefMagnitudeF= 45
19   res@ vcRefLengthF= 0.08
20   res@ vcRefAnnoPerimOn= True
21   res@ vcRefAnnoString1On= True
22   res@ vcRefAnnoString1= "45"
23   res@ vcRefAnnoString2On= True
```

```
24   res@ vcRefAnnoString2= "m/s"
25   res@ vcRefAnnoBackgroundColor= "White"
26   res@ vcRefAnnoOrthogonalPosF= - 1.0
27   res@ vcRefAnnoParallelPosF= 1.0
28   res@ vcRefAnnoArrowLineColor= "black"
29   res@ vcRefAnnoFontHeightF= 0.02
30   plot= gsn_csm_vector_map(wks,u(::5,::5),v(::5,::5),res)
31   end
```

Line 11 sets to draw vectors last.

Line 12 sets to turn on curly vectors.

Line 13 sets to a small value to thin the vectors.

Line 14 draws the minimum value of vectors.

Line 15 changes the vector color.

Line 16 changes the vector thickness.

Line 17 draws the reference vectors.

Line 18 defines the vector reference magnitude.

Line 19 defines the length of the reference vector.

Line 20 draws the border of the reference vector.

Line 21 draws a string above the reference vector.

Line 22 draws the string above the reference vector.

Line 23 draws a string under the reference vector.

Line 24 draws the string under the reference vector.

Line 25 sets the background color to fill the box surrounding the vector reference annotation.

Line 26 moves the reference vector vertically.

Line 27 moves the reference vector horizontally.

Line 28 changes the reference vector color.

Line 29 changes the font size of the reference vector label.

Results:

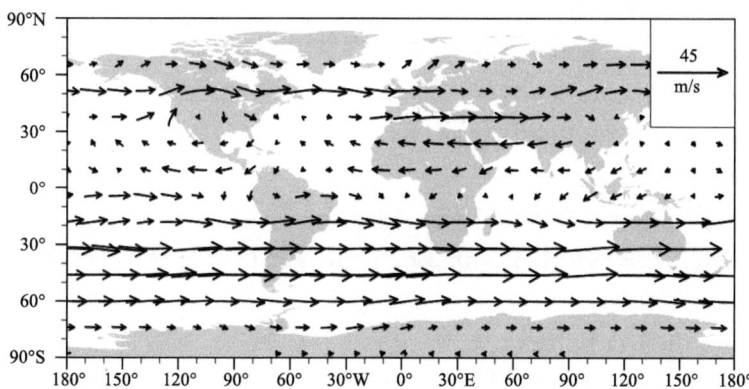

Example 7-2-4:

Do the same as you did in Example 7-2-3, but draw a vector plot over a polar stereographic map.

Code:

```
1    begin
2    a= addfile("./uv300.nc","r")
3    u= a-> U(1,:,:)
4    v= a-> V(1,:,:)
5    wks= gsn_open_wks("pdf","ex7.2.4")
6    res= True
7    res@ gsnLeftString = ""
8    res@ gsnRightString= ""
9    res@ gsnPolar= "NH"
10   res@ vcMinDistanceF= 0.04
11   res@ vcMinMagnitudeF= 1.0
12   res@ vcPositionMode= "ArrowTail"
13   res@ vcRefAnnoOn= True
14   res@ vcRefMagnitudeF= 15
15   res@ vcRefLengthF= 0.06
16   res@ vcRefAnnoPerimOn= False
17   res@ vcRefAnnoString1On= True
18   res@ vcRefAnnoString1= "15"
19   res@ vcRefAnnoString2On= True
20   res@ vcRefAnnoString2= "m/s"
21   res@ vcRefAnnoBackgroundColor= -1
22   res@ vcRefAnnoOrthogonalPosF= -1.0
23   res@ vcRefAnnoParallelPosF= 0.1
24   res@ vcRefAnnoFontHeightF= 0.02
25   res@ vcGlyphStyle= "FillArrow"
26   res@ vcFillArrowWidthF= 0.07
27   res@ vcFillArrowHeadXF= 0.6
28   res@ vcFillArrowHeadYF= 0.25
29   res@ vcFillArrowHeadInteriorXF= 0.25
30   res@ vcFillArrowFillColor= "blue"
31   res@ vcFillArrowEdgeColor= "White"
32   plot = gsn_csm_vector_map_polar(wks,u(::4,::4),v(::4,::4),res)
33   end
```

Line 9 chooses the hemisphere.

Line 12 sets the position of the lattice corresponds to the end of the arrow.

Line 21 indicates that the background color of the reference arrow is set to be transparent.

Line 25—31 set the style of the fill arrows.

Line 32 creates and draws a vector plot over a polar stereographic map.

Results:

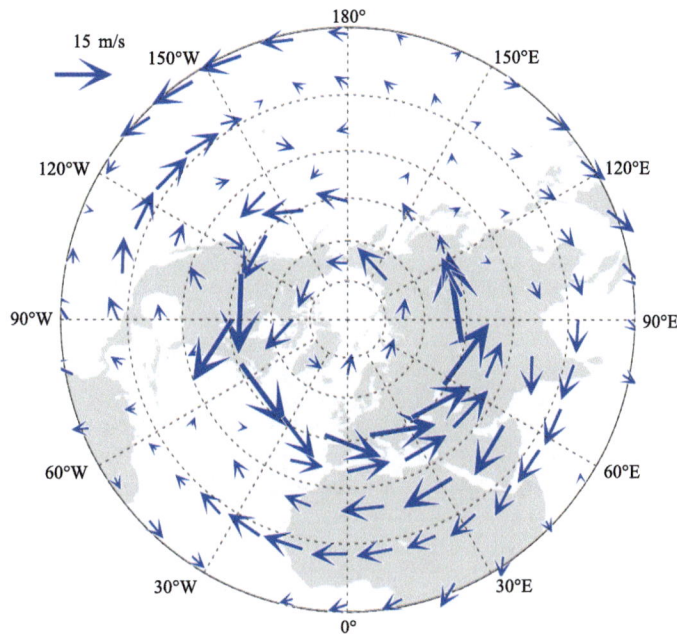

Example 7-2-5:

Do the same as you did in Example 7-2-4, but set color vectors based on magnitude.

Code:

```
1   Begin
2   a= addfile("./uv300.nc","r")
3   u= a- > U(1,:,:)
4   v= a- > V(1,:,:)
5   wks= gsn_open_wks("pdf","ex7.2.5")
6   cmap= read_colormap_file("BlAqGrYeOrReVi200")
7   res= True
8   res@ gsnPolar= "NH"
9   res@ vcLevelPalette= cmap(15:180,:)
10  res@ vcGlyphStyle= "CurlyVector"
11  res@ vcLineArrowThicknessF= 2.5
12  res@ vcRefMagnitudeF= 15
13  res@ vcRefLengthF= 0.06
14  res@ vcMonoLineArrowColor= False
15  plot= gsn_csm_vector_map_polar(wks,u(::4,::4),v(::4,::4),res)
```

16   end

Line 6 is to choose a color map.

Line 9 subsets the color map.

Line 14 sets vectors' colors by magnitude.

Results:

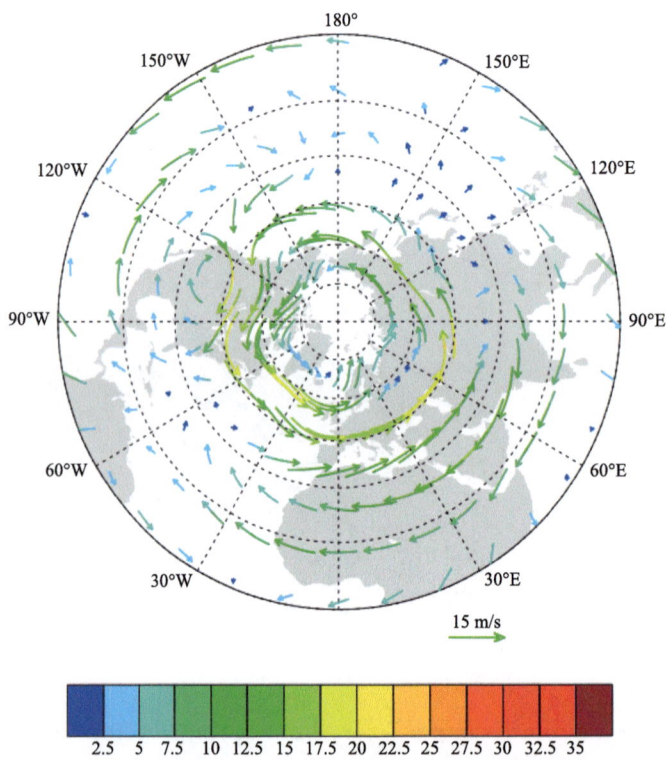

> **Practices**

Draw vectors for the following questions. The relevant data used in the following question comes from this web site: https://psl. noaa. gov/data/gridded/data. ncep. reanalysis2. html

(1) Draw the summer wind field at 300hPa in 70°S—70°N. Set the style of the fill arrows and the reference vector.

(2) Draw a global summer wind field at 300 hPa over a polar stereographic map. Set vectors' colors based on magnitude.

7.3 Overlay plot

In NCL, Overlay plot is a way to Overlay multiple layers together to create a composite graph. It should be noted that if the two maps to be superimposed contain a map, the map can only be drawn in the base map. —*Overlay plot*.

➢ Key points

(1) Overlay one plot object on another

$$\text{overlay(base\_id, transform\_id)}$$

The arguments used in the function are set as follows:

• *base_id*: The base plot of the overlay chain of plots. The object referenced must support overlays.

• *transform_id*: The graphical object is to be overlaid on the *base_id* plot. The object must be one of the Transform objects; that is, an object that can go through a transformation to become part of the *base_id* plot. Examples of Transform objects include contour plots, XY plots, vector plots, and streamline plots.

(2) Create and draw two contour plots over a map

$$\text{gsn\_csm\_contour\_map\_overlay(wks, data1, data2, res1, res2)}$$

The arguments used in the function are set as follows:

• *wks*: A workstation identifier. The identifier is one returned either from calling **gsn_open_wks** or calling create to create a workstation object.

• *data1*: The first set of data to contour; must be one-or two-dimensional.

• *data2*: The second set of data to contour; must be one-or two-dimensional.

• *res1*: A variable containing an optional list of plot resources, attached as attributes, is to be applied to the first set of contour data and/or the map plot. Set this value to **True** if you want the attached attributes to be applied, and **False** if you either don't have any resources to set, or you don't want the resources applied.

• *res2*: A variable containing an optional list of plot resources, attached as attributes, is to be applied to the second set of contour data. Set this value to **True** if you want the attached attributes to be applied, and **False** if you either don't have any resources to set, or you don't want the resources applied. Note that this resource is for the second contour plot only, and should not have any map ("*mp*") resources attached to it.

➢ Examples

Example 7-3-1:

Overlay line plots. The data source is the same as Example 7-1-1.

Code:

```
1   begin
2   f= addfile ("./uv300.nc","r")
3   lat= f- > lat
4   u= f- > U
5   lon= (/60., - 45., 0./)
6   u0= u(0,:,{lon(0)})
7   u1= u(0,:,{lon(1)})
8   u2= u(0,:,{lon(2)})
9   colors= (/"red","green","blue"/)
```

```
10  wks= gsn_open_wks("pdf","ex7.3.1")
11  res= True
12  res@ gsnDraw= False
13  res@ gsnFrame= False
14  res@ xyLineThicknessF= 4.0
15  res@ xyLineColor= colors(0)
16  plot0= gsn_csm_xy(wks,lat,u0,res)
17  res@ xyLineColor= colors(1)
18  plot1= gsn_csm_xy(wks,lat,u1,res)
19  res@ xyLineColor= colors(2)
20  plot2= gsn_csm_xy(wks,lat(5:),u2(5:),res)
21  overlay(plot0,plot1)
22  overlay(plot0,plot2)
23  lgres= True
24  lgres@ lgLineColors= colors
25  lgres@ lgItemType= "Lines"
26  lgres@ lgLabelFontHeightF = .1
27  lgres@ vpWidthF= 0.13
28  lgres@ vpHeightF= 0.10
29  lgres@ lgPerimThicknessF= 3.0
30  lgres@ lgMonoDashIndex= True
31  lgres@ lgDashIndex= 0
32  labels= "lon= "+ lon
33  legend= gsn_create_legend (wks, 3, labels,lgres)
34  amres= True
35  amres@ amJust= "BottomRight"
36  amres@ amParallelPosF= 0.5
37  amres@ amOrthogonalPosF= 0.5
38  annoid= gsn_add_annotation(plot0,legend,amres)
39  draw(plot0)
40  frame(wks)
41  end
```

Line 21 overlays plot1 onto plot0.

Line 22 overlays plot2 onto plot0.

Line 23 attaches a legend.

Line 24 sets the color assigned to each legend item line.

Line 25 shows lines only.

Line 26 changes the font thickness of the legend label.

Line 27 changes the width of the legend.

Line 28 changes the height of the legend.

Line 29 thickens the box perimeter.

Line 30—31 set a uniform solid pattern index for the lines drawn in all items in the legend.

Line 32—33 create a legend.

Line 35 uses the bottom right corner of box.

Line 36 moves the legend right.

Line 37 moves the legend down.

Line 38 adds the legend to the plot.

Results:

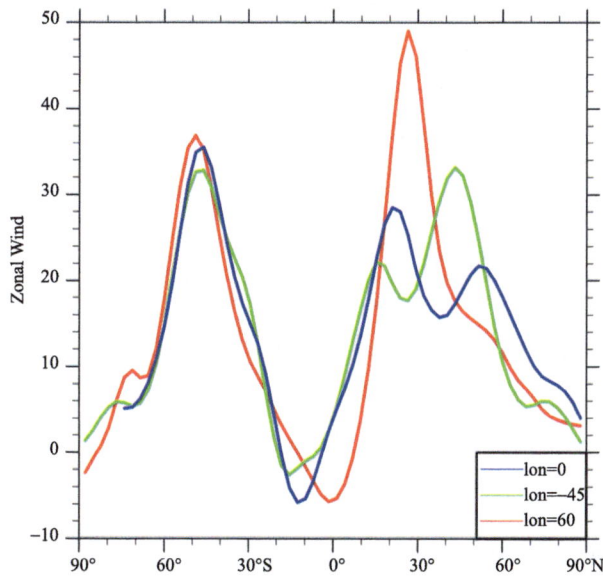

Example 7-3-2:

Draw the u wind and sea surface temperature (SST) data together with overlaying line contours for wind on filled contours for SST. The wind field data used in the following examples comes from this web site: https://psl.noaa.gov/data/gridded/data.ncep.reanalysis2.html. The SST data comes from this web site: https://www.ncei.noaa.gov/pub/data/cmb/ersst/v5/netcdf.

Code(way 1):

```
1    begin
2    f1= addfile ("./uwnd.mon.mean197901- 202104- grid1.nc","r")
3    f2= addfile ("./sst_ERSSTv5_197901_202104_grid1.nc","r")
4    uwnd= f1- > uwnd(0,0,:,:)
5    temp= f2- > sst(0,:,:)
6    wks= gsn_open_wks("pdf","ex7.3.2.1")
7    res= True
```

```
 8    res@ gsnDraw= False
 9    res@ gsnFrame= False
10    res@ mpFillOn= False
11    res@ mpCenterLonF= 180.0
12    res@ mpMaxLatF= 60.0
13    res@ mpMinLatF= 20.0
14    res@ mpMinLonF= 150.0
15    res@ mpMaxLonF= 210.0
16    res@ cnLevelSelectionMode= "ExplicitLevels"
17    res@ cnLevels= ispan(-30,30,5)
18    res@ cnLineLabelsOn= False
19    res@ cnFillOn= True
20    res@ cnLinesOn= False
21    res@ cnFillPalette= "BlueDarkRed18"
22    sres= True
23    sres@ gsnDraw= False
24    sres@ gsnFrame= False
25    plot= gsn_csm_contour_map(wks,temp,res)
26    plot_ov= gsn_csm_contour(wks,uwnd,sres)
27    overlay(plot,plot_ov)
28    draw(plot)
29    frame(wks)
30    end
```

Line 11—15 specify the plot domain.

Line 16 uses explicit levels.

Line 17 sets the contour levels.

Line 18 sets not to use line labels.

Line 19 sets to fill with color.

Line 20 sets not to draw contour lines.

Line 22—24 set up a second resource list.

Line 25 creates the temperature plot.

Line 26 creates the U-wind plot.

Line 27 overlays the U-wind plot on the temperature plot.

Line 28 draws the temperature plot (with the U-wind plot overlaid).

Results:

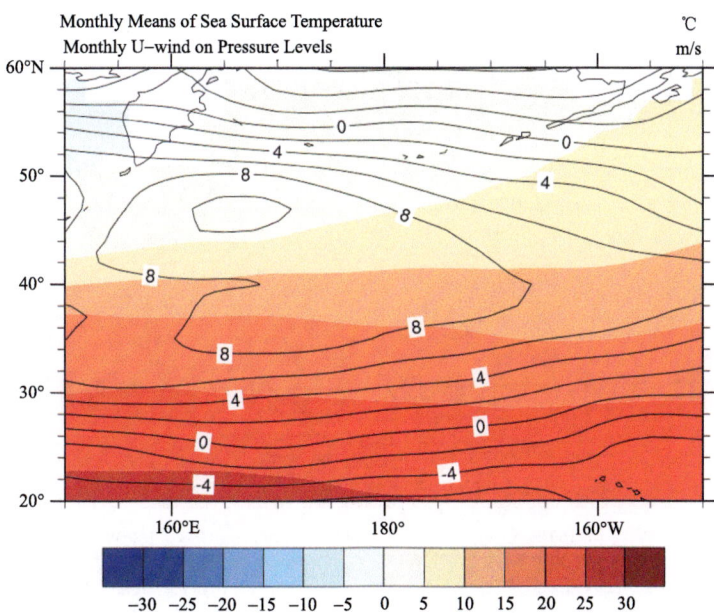

Code(way 2):

```
1   Begin
2   f1= addfile ("./uwnd.mon.mean197901- 202104- grid1.nc","r")
3   f2= addfile ("./sst_ERSSTv5_197901_202104_grid1.nc","r")
4   uwnd= f1- > uwnd(0,0,:,:)
5   temp= f2- > sst(0,:,:)
6   wks= gsn_open_wks("pdf","ex7.3.2.2")
7   res= True
8   res@ mpFillOn= False
9   res@ mpCenterLonF= 180.
10  res@ mpMaxLatF= 60.
11  res@ mpMinLatF= 20.
12  res@ mpMinLonF= 150.
13  res@ mpMaxLonF= 210.
14  res@ cnLevelSelectionMode= "ExplicitLevels"
15  res@ cnLevels= ispan(- 30,30,5)
16  res@ cnLineLabelsOn= False
17  res@ cnFillOn= True
18  res@ cnLinesOn= False
19  res@ cnFillPalette= "BlueDarkRed18"
20  sres= True
21  plot= gsn_csm_contour_map_overlay(wks,temp,uwnd,res,sres)
22  end
```

Line 21 creates the overlay plot.
Results:

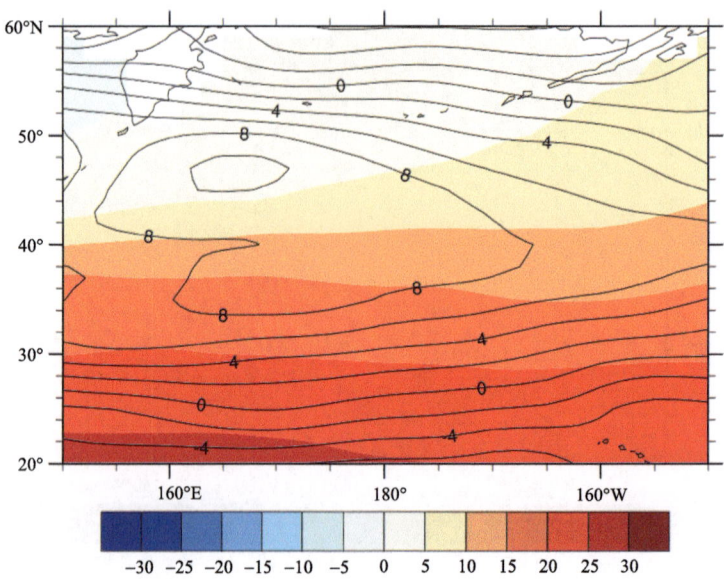

Example 7-3-3:

The overlay of contours and vectors. The data source is the same as Example 7-3-2.

Code:

```
1   begin
2   f1= addfile ("./uwnd.mon.mean197901- 202104- grid1.nc","r")
3   f2= addfile ("./vwnd.mon.mean197901- 202104- grid1.nc","r")
4   f3= addfile ("./sst_ERSSTv5_197901_202104_grid1.nc","r")
5   u= f1- > uwnd(0,2,:,:)
6   v= f2- > vwnd(0,2,:,:)
7   t= f3- > sst(0,:,:)
8   wks= gsn_open_wks("pdf","ex7.3.3")
9   res= True
10  res@ gsnFrame= False
11  res@ tmXTOn= False
12  res@ tmYROn= False
13  res@ gsnLeftString  = ""
14  res@ gsnRightString= ""
15  res@ mpFillOn= False
16  res@ mpCenterLonF= 180.
17  res@ mpMinLatF= - 10.0
18  res@ mpMaxLatF= 60.0
19  res@ mpMinLonF= 120.0
20  res@ mpMaxLonF= 240.0
```

```
21    res@ cnFillOn= True
22    res@ cnFillPalette = "testcmap"
23    res@ cnLinesOn  = False
24    resv= True
25    resv@ gsnDraw= False
26    resv@ gsnFrame= False
27    resv@ gsnLeftString= ""
28    resv@ gsnRightString= ""
29    resv@ vcMinDistanceF= 0.025
30    resv@ vcLineArrowThicknessF= 2
31    resv@ vcGlyphStyle= "LineArrow"
32    resv@ vcRefAnnoOn= True
33    resv@ vcRefMagnitudeF= 20
34    resv@ vcRefAnnoString1On= True
35    resv@ vcRefAnnoString1= "20m/s"
36    resv@ vcRefAnnoSide=  "Top"
37    resv@ vcRefAnnoString2On= False
38    resv@ vcRefAnnoPerimOn= True
39    resv@ vcRefAnnoOrthogonalPosF= - 0.25
40    resv@ vcRefAnnoParallelPosF= 1
41    resv@ vcRefLengthF= 0.1
42    resv@ vcRefAnnoBackgroundColor= "White"
43    resv@ vcRefAnnoFontHeightF= 0.02
44    resv@ vcVectorDrawOrder= "PostDraw"
45    resv@ gsnStringFontHeightF= 0.035
46    contour= gsn_csm_contour_map(wks, t, res)
47    vector= gsn_csm_vector(wks,u, v, resv)
48    overlay(contour,vector)
49    draw(contour)
50    frame(wks)
51    end
```

Line 15—20 set the area.

Line 21—23 set the contour.

Line 24—45 set the vector.

Line 46—47 create plots.

Line 48 overlays the vector plot on the contour plot.

Results:

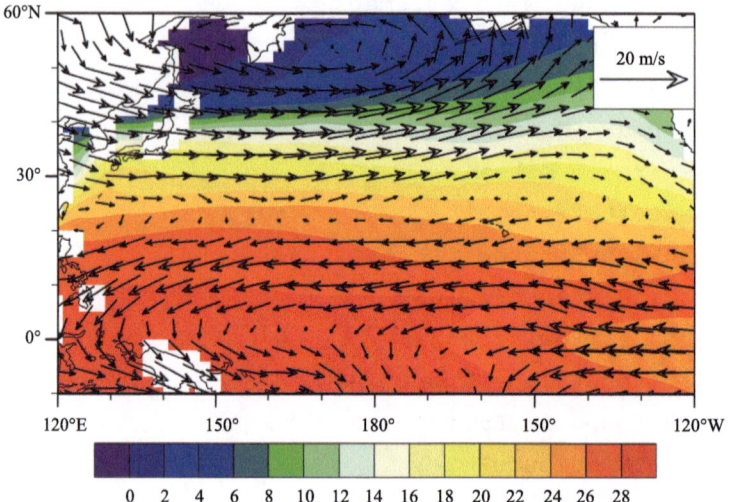

➢ **Practices**

Draw overlay plots for the following questions. The wind field data used in the following question comes from this web site: https://psl.noaa.gov/data/gridded/data.ncep.reanalysis2.html. The SST data comes from this web site: https://www.ncei.noaa.gov/pub/data/cmb/ersst/v5/netcdf.

(1) Draw the Line plot overlay: the summer zonal winds at 850 hPa over the equator and the SST in the equator during the 1979—2021.

(2) Draw the overlay of contours and vectors: the summer wind field at 850 hPa and the SST field in the range of 60°S—60°N.

Chapter 8　Synthesized Example

8.1　Spectral analysis

Spectral analysis is a method used to analyze the frequency components of time series signals. This method usually transforms the signal into the time domain and the frequency domain, which makes it easier to carry on the quantitative analysis of the signal characteristic. The latter is called the spectrum. The spectrum breaks the sample variance of time series into discrete components, each of which is associated with a particular frequency.

➢ **Key points**

(1) Calculate the spectra of a series

$$\text{specx\_anal(x,iopt,jave,pct)}$$

The arguments used in the function are set as follows:

• x: A one-dimensional array containing the data. Missing values are not allowed.

• $iopt$: A scalar representing the detrending option. If $iopt=0$, remove series mean. If $iopt=1$, remove the series mean and least squares linear trend.

• $jave$: A scalar representing the smoothing to be performed on the periodogram estimates. This should be an odd number ($\geqslant 3$). If not, the routine will force it to the next largest odd number. If $jave<3$, do no smoothing. $spcx$ contains raw spectra estimates (periodogram). If $jave\geqslant 3$, average $jave$ periodogram estimates together utilizing modified Daniell smoothing (good stability but may lead to large bias). All weights are $1/jave$ except weight(1) and weight($jave$) which are $1/(2\times jave)$. This is the recommended option. It is this number which has the most impact on the degrees of freedom.

• pct: A scalar representing the percent of the series to be tapered ($0.0 \leqslant pct \leqslant 1.0$). If $pct=0.0$, no tapering will be done. If $pct=1.0$, the whole series is affected. A value of 0.10 is common (tapering should always be done).

(2) Calculate cross spectra quantities of a series

$$\text{specxy\_anal(x,y,iopt,jave,pct)}$$

The arguments used in the function are set as follows:

• x/y: One-dimensional arrays containing the data. x and y must be the same length, and missing values are not allowed.

• The rest of the arguments in the function are set the same as above.

(3) Calculate the theoretical Markov spectrum and the lower and upper

confidence curves

$$\text{specx\_ci(sdof,lowval,highval)}$$

The arguments used in the function are set as follows:

• *sdof*: A degrees-of-freedom array returned from the NCL functions *specx_anal* or *specxy_anal*.

• *lowval*: The lower confidence limit ($0.0 < lowval < 1.$). A typical value is 0.05.

• *Highval*: The upper confidence limit ($0.0 < hival < 1.$). A typical value is 0.95.

➢ **Examples**

Example 8-1-1:

Perform a spectral analysis on series x. The SST data comes from this web site: https://www.ncei.noaa.gov/pub/data/cmb/ersst/v5/netcdf.

Code:

```
1   begin
2     f= addfile("./ersst.v5.1979- 2016.nc","r")
3     sst= f- > sst(:,0,{- 5:5},{190:240})
4     sst_avg= clmMonTLL(sst)
5     ssta= calcMonAnomTLL(sst, sst_avg)
6     s_index= dim_avg_n_Wrap(ssta, (/1,2/))
7     iopt= 0
8     jave= 7
9     pct= 0.1
10    spec= specx_anal (s_index,iopt,jave,pct)
11    wks= gsn_open_wks("pdf","ex8.1.1")
12    res= True
13    res@ tiMainString= "Nino3.4"
14    res@ tiXAxisString= "Frequency (cycles/month)"
15    res@ tiYAxisString= "Variance"
16    plot= gsn_csm_xy(wks,spec@ frq,spec@ spcx,res)
17  end
```

Line 1 is the beginning of program.

Line 2—3 open the file and read in data.

Line 4—6 calculate the Niño3.4 index.

Line 7—9 set the function arguments.

Line 7 removes the series mean.

Line 8 sets the smoothing periodogram.

Line 9 tapers 10% of the data.

Line 10 calculates the spectrum.

Line 13 sets the title.

Line 14 sets the x-axis.

Line 15 sets the y-axis.
Line 16 creates the plot.
Results:

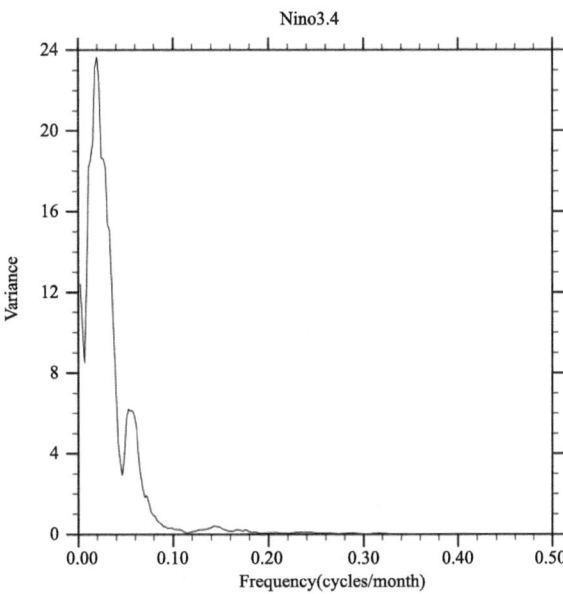

Example 8-1-2:

Read two time series and calculate their spectra, cospectra, coherence, quadrature spectra, and phase and create a panel plot of all values. The data used in the following example comes from this web site: https://www.ncl.ucar.edu/Applications/Data.

Code:

```
1    begin
2    f= addfile("./SLP_DarwinTahitiAnom.nc","r")
3    dslp= f-> DSLP
4    tslp= f-> TSLP
5    d= 0
6    sm= 7
7    pct= 0.10
8    spec= specxy_anal(dslp,tslp,d,sm,pct)
9    plot= new(6,graphic)
10   wks= gsn_open_wks("pdf","ex8.1.2")
11   res= True
12   res@ tiXAxisString= "Frequency (cycles/month)"
13   res@ gsnFrame= False
14   res@ gsnDraw= False
15   res@ tiYAxisString= "Cospectrum"
16   plot(0)= gsn_csm_xy(wks,spec@ frq,spec@ cospc,res)
```

```
17  res@ tiYAxisString= "Quadrature"
18  plot(1)= gsn_csm_xy(wks,spec@ frq,spec@ quspc,res)
19  res@ tiYAxisString= "Coherence SQ"
20  plot(2)= gsn_csm_xy(wks,spec@ frq,spec@ coher,res)
21  res@ tiYAxisString= "Phase"
22  plot(3)= gsn_csm_xy(wks,spec@ frq,spec@ phase,res)
23  res@ tiYAxisString= "Variance of DSLP"
24  plot(4)= gsn_csm_xy(wks,spec@ frq,spec@ spcx,res)
25  res@ tiYAxisString= "Variance of TSLP"
26  plot(5)= gsn_csm_xy(wks,spec@ frq,spec@ spcy,res)
27  res_P= True
28  res_P@ gsnMaximize= True
29  gsn_panel(wks,plot,(/3,2/),res_P)
30  end
```

Line 1 is the beginning of program.

Line 2—4 open the file and read in data.

Line 5—7 set the function arguments.

Line 5 removes the series mean.

Line 6 sets the smoothing periodogram.

Line 7 tapers 10% of the data.

Line 8 calculates the spectrum.

Line 9—14 are plotting parameters that remain constant.

Line 15—16 create a plot of cospectrum.

Line 17—18 create a plot of quadrature spectrum.

Line 19—20 create a plot of coherence.

Line 21—22 create a plot of phase.

Line 23—24 create a plot spec soi.

Line 25—26 create a plot spec slp.

Line 27 sets the desired panel mods.

Line 28 blows up the plot.

Line 29 creates the panel plot.

Results:

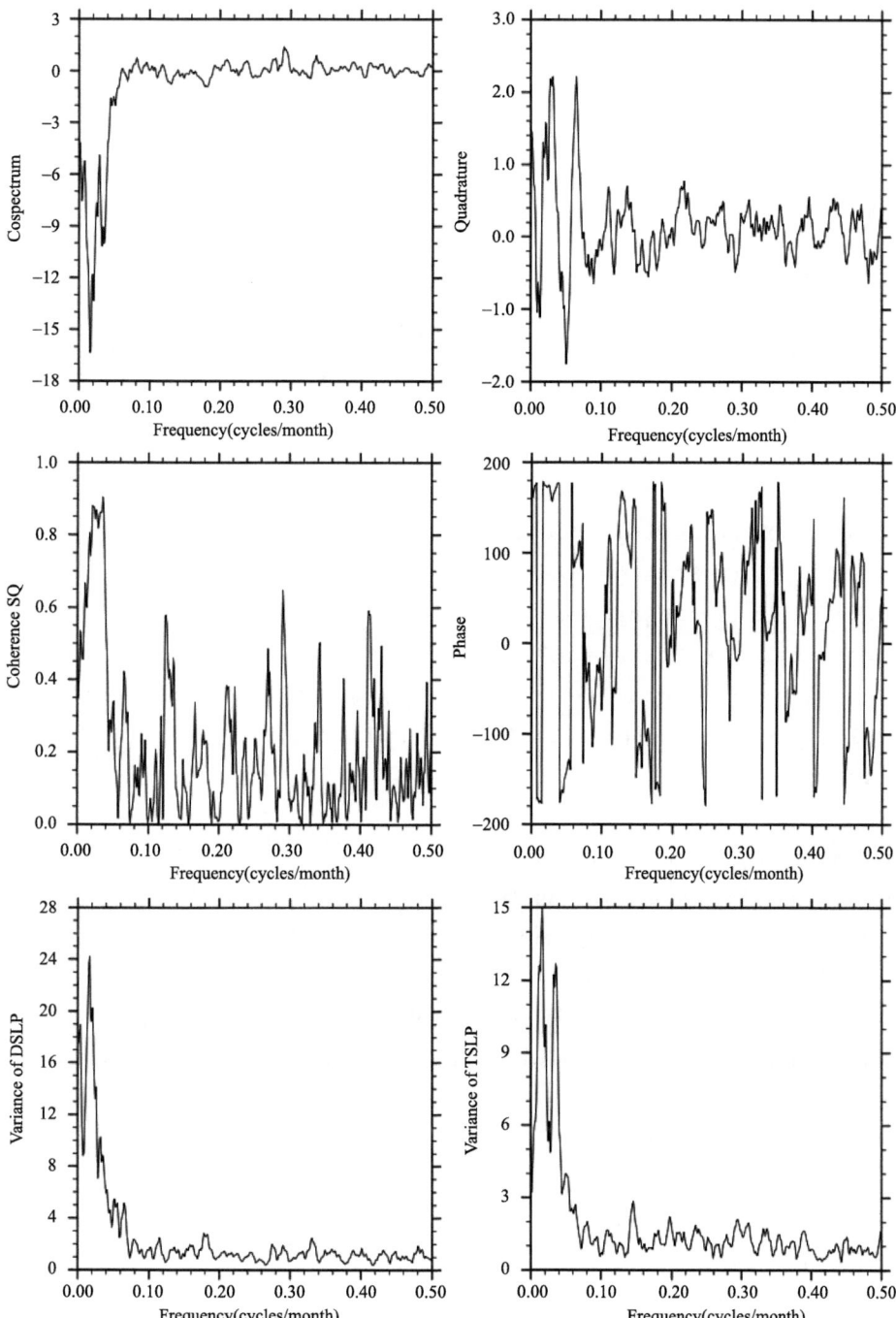

Example 8-1-3:

Calculate the confidence intervals. The data used in the following example comes from this web site: https://www.ncl.ucar.edu/Applications/Data.

Code:

```
1   begin
2   f= addfile("./SOI_Darwin.nc","r")
3   soi= f- > DSOI
4   d= 0
5   sm= 21
6   pct= 0.10
7   sdof= specx_anal(soi,d,sm,pct)
8   splt= specx_ci(sdof, 0.05, 0.95)
9   wks= gsn_open_wks("pdf","ex8.1.3")
10  res= True
11  res@ gsnDraw= False
12  res@ gsnFrame= False
13  res@ tiMainString= "SOI"
14  res@ tiXAxisString= "Frequency (cycles/month)"
15  res@ tiYAxisString= "Variance"
16  res@ xyLineThicknesses= (/2.,1.,1.,1./)
17  res@ xyDashPatterns= (/0,0,1,1/)
18  res@ xyLineColors= (/"foreground","green","blue","red"/)
19  plot= gsn_csm_xy(wks,sdof@ frq, splt,res)
20  draw(plot)
21  end
```

Line 1 is the beginning of program.

Line 2—3 open the file and read in data.

Line 4—6 set the function arguments.

Line 7 calculates the spectrum.

Line 8 calculates the confidence intervals. splt(0,:)-input spectrum; splt(1,:)-Markov "Red Noise" spectrum; splt(2,:)-lower confidence bound for Markov; splt(3,:)-upper confidence bound for Markov.

Line 9—18 set plot parameters.

Line 17—18 create the plot of quadrature spectrum.

Line 19 creates the plot. sdof@frq: a one-dimensional array of length $N/2$ representing frequency (cycles/time).

Results:

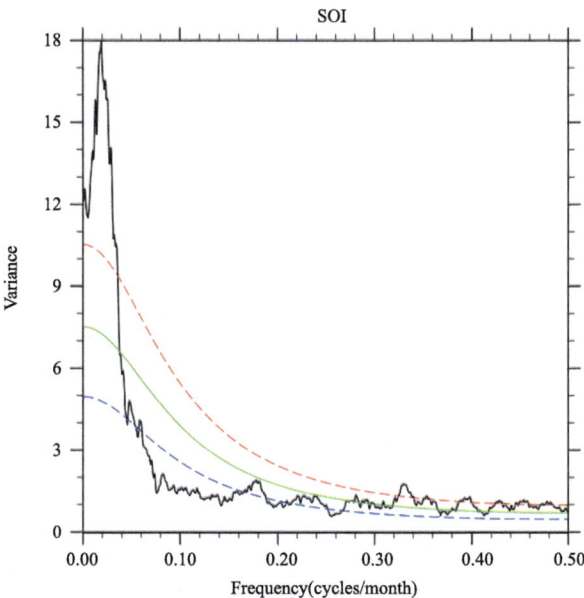

Example 8-1-4:

Plot spectra and "red noise" confidence intervals on a log scale. The data is the same as Example 8-1-3.

Code:

```
1   begin
2   f= addfile("./SOI_Darwin.nc","r")
3   soi= f-> DSOI
4   d= 0
5   sm= 21
6   pct= 0.10
7   sdof= specx_anal(soi,d,sm,pct)
8   splt= specx_ci (sdof, 0.05, 0.95)
9   wks= gsn_open_wks("pdf","ex8.1.4")
10  res= True
11  res@ tiMainString= "SOI"
12  res@ tiXAxisString= "Frequency (cycles/month)"
13  res@ tiYAxisString= "Variance"
14  res@ trYLog= True
15  res@ trYMinF= 0.10
16  res@ trYMaxF= 30.0
17  res@ gsnFrame= False
18  plot= gsn_csm_xy(wks,sdof@ frq, splt,r)
```

```
19    end
```
Line 4—6 set the function arguments.
Line 7 calculates the spectrum.
Line 8 calculates the confidence intervals.
Line 14 sets the log scaling.
Line 15 sets the lower limit manually.
Line 16 sets the upper limit manually.
Results:

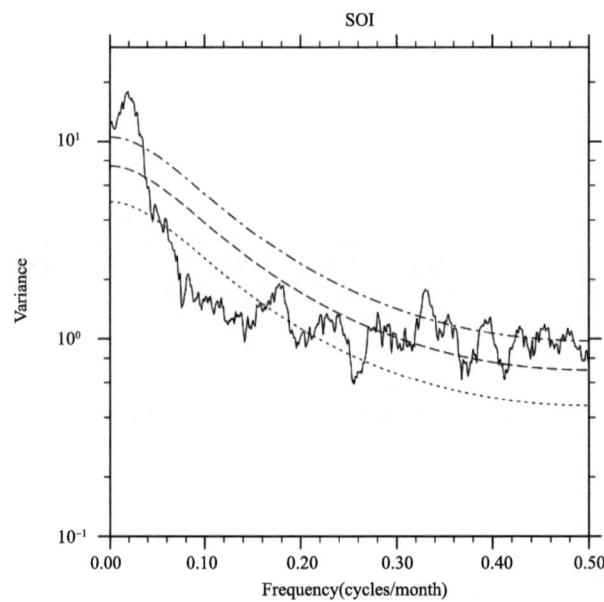

Example 8-1-5:
Use period (1/fequency) as the abscissa. The data is the same as Example 8-1-3.
Code(way 1):
```
1     begin
2     f= addfile("./SOI_Darwin.nc","r")
3     soi= f->DSOI
4     d= 0
5     sm= 21
6     pct= 0.10
7     sdof= specx_anal(soi,d,sm,pct)
8     splt= specx_ci(sdof,0.05,0.95)
9     wks= gsn_open_wks("pdf","ex8.1.5")
10    res= True
11    res@ tiMainString= "SOI: Darwin"
12    res@ tiXAxisString= "Period (month)"
13    res@ tiYAxisString= "Variance/freq"
```

```
14   res@ trYMinF= 0.00
15   res@ trYMaxF= 20.0
16   f= sdof@ frq
17   p= 1/f
18   p!0= "f"
19   p&f= f
20   p@ long_name= "period"
21   p@ units= "month"
22   ip= ind(p.le.240)
23   plot= gsn_csm_xy(wks,p(ip), splt(:,ip),res)
24   end
```

Line 17—23 plot abscissa as period (1/frequency).

Results:

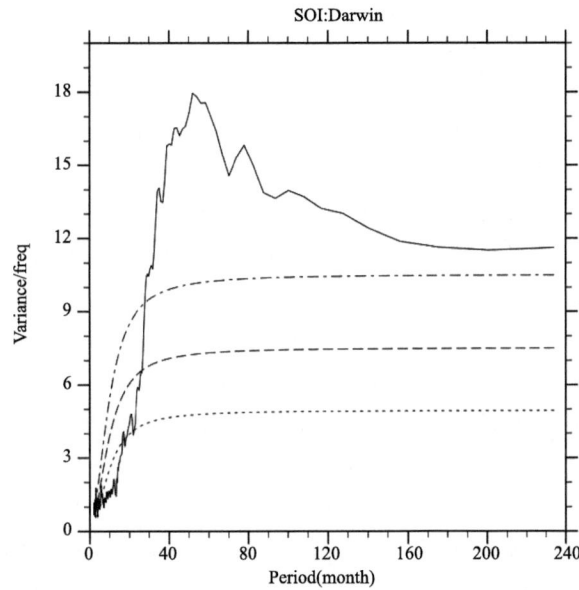

> **Practices**

Calculate the spectrum and plot the results for the following questions. The Nino3.4 index comes from https://climatedataguide.ucar.edu/climate-data/nino-sst-indices-nino-12-3-34-4-oni-and-tni.

(1) Plot winter Niño3.4 index spectra and calculate confidence intervals.

(2) Plot summer Niño3.4 index spectra and calculate confidence intervals, using period (1/frequency) as the abscissa.

8.2 EOF analysis

EOF stands for Empirical Orthogonal Functions (aka: Principal Component Analysis,

Eigen Analysis). It is a statistical technique commonly used in geophysical and atmospheric sciences to analyze and represent the variability of spatial datasets in terms of their dominant patterns or modes of variability. EOF analysis involves decomposing a dataset into a set of orthogonal functions that capture as much variance as possible, with the first few EOFs typically explaining most of the total variance. Each EOF represents a distinct pattern or structure from which one can discern important characteristics about the underlying physical process generating that particular dataset. These analyses are widely used for climate research, weather forecasting, oceanography, and other related areas where large multivariate data sets must be reduced or interpreted for practical applications.

➢ **Key points**

(1) Empirical orthogonal functions

eofunc_n_Wrap(data,neval,optEOF,dim)

The arguments used in the function are set as follows:

• *data*: A multi-dimensional array in which the dim index specifies the dimension that contains the number of observations.

• *neval*: A scalar integer that specifies the number of eigenvalues and eigenvectors to be returned. This is usually less than or equal to the minimum number of observations or number of variables, and is typically 3 to 5.

• *optEOF*: A logical variable to which various optional arguments may be assigned as attributes. These optional arguments alter the default behavior of the function. Must be set to **True** prior to setting the attributes which are assigned using the @ operator.

• *dim*: The dimension index of data that represents the dimension containing the number of observations. Generally, this is the time dimension.

(2) Calculate the time series of the amplitudes associated with each eigenvalue in an EOF

eofunc_ts_n_Wrap(data,evec,optETS,dim)

The arguments used in the function are set as follows:

• *data*: A multi-dimensional array in which the dim index specifies the dimension that contains the number of observations. Generally, this is the time dimension.

• *evec*: A multi-dimensional array containing the EOFs calculated using **eofunc_n** or **eofunc_n_Wrap**.

• *optETS*: A logical variable to which various optional arguments may be assigned as attributes. These optional arguments alter the default behavior of the function. Must be set to **True** prior to setting the attributes which are assigned using the @ operator.

• *dim*: The dimension index of data that represents the dimension containing the number of observations. Generally, this is the time dimension.

➢ **Examples**

Example 8-2-1:

EOF analysis is performed for SST anomalies. The data used in the following example come from this web site: https://www.metoffice.gov.uk/hadobs/hadisst/data/download.html.

Code:

```
1   begin
2   a= addfile("./anom_detrend_sst_HadISST_198001- 202212.nc", "r")
3   sst= a- > sst(:,{- 30:30},{120:270})
4   neof= 4
5   eof= eofunc_n_Wrap(sst, neof, False, 0)
6   eof= eof* - 1.
7   eof_ts= eofunc_ts_n_Wrap (sst, eof, False, 0)
8   eof_ts_std= dim_standardize_n(eof_ts, 0, 1)
9   do ieof= 1,neof
10  eof(ieof- 1,:,:)= eof(ieof- 1,:,:) * stddev(eof_ts(ieof- 1,:))
11  end do
12  wks= gsn_open_wks("pdf","ex8.2.1")
13  gsn_define_colormap(wks, "BlueYellowRed")
14  res= True
15  res@ gsnDraw= False
16  res@ gsnFrame= False
17  res@ mpShapeMode= "FreeAspect"
18  res@ vpWidthF= 0.7
19  res@ vpHeightF= 0.4
20  res@ tiYAxisString= ""
21  res@ gsnAddCyclic= False
22  res@ gsnLeftStringFontHeightF= 0.035
23  res@ gsnRightStringFontHeightF= 0.035
24  res@ tmXTOn= False
25  res@ tmYROn= False
26  res@ cnFillOn= True
27  res@ cnLinesOn= False
28  res@ lbLabelBarOn= True
29  res@ lbOrientation= "vertical"
30  res@ cnLevelSelectionMode= "ManualLevels"
31  res@ cnMinLevelValF= - 0.5
32  res@ cnMaxLevelValF= 0.5
33  res@ cnLevelSpacingF= 0.1
34  res@ mpMinLatF= - 30
35  res@ mpMaxLatF= 30
36  res@ mpMinLonF= 120
37  res@ mpMaxLonF= 270
38  res@ mpCenterLonF= 180
```

```
39   res@ tmXBLabelFontHeightF= 0.025
40   res@ tmYLLabelFontHeightF= 0.025
41   xyres= True
42   xyres@ gsnDraw= False
43   xyres@ gsnFrame= False
44   xyres@ vpWidthF= 0.9
45   xyres@ vpHeightF= 0.35
46   xyres@ gsnLeftStringFontHeightF= 0.035
47   xyres@ gsnRightStringFontHeightF= 0.035
48   xyres@ tmXTOn= False
49   xyres@ tmYROn= False
50   xyres@ gsnYRefLine= 0.0
51   xyres@ trYMinF= - 4.0
52   xyres@ trYMaxF= 4.0
53   xyres@ tmXBMode= "Explicit"
54   xyres@ tmXBLabelAngleF= - 45.0
55   xyres@ tmXBLabelJust= "CenterLeft"
56   xyres@ tmXBValues= ispan(1,dimsizes(eof_ts_std(0,:)),5* 12)
57   xyres@ tmXBLabels= ispan(1980,2022,5)
58   xyres@ tmXBMinorValues= ispan(1,dimsizes(eof_ts_std(0,:)),12)
59   xyres@ tmXBLabelFontHeightF= 0.025
60   xyres@ tmYLLabelFontHeightF= 0.025
61   plot= new(8, graphic)
62   do ieof= 1,neof
63   res@ gsnLeftString= "EOF"+ ieof
64   res@ gsnRightString= sprintf("% 5.1f", eof@ pcvar(ieof- 1)) + "% "
65   plot((ieof- 1)* 2+ 0)= gsn_csm_contour_map(wks, eof(ieof- 1,:,:),
       res)
66   xyres@ gsnLeftString= "PC"+ ieof
67   plot((ieof- 1)* 2+ 1)= gsn_csm_xy(wks,ispan(1,dimsizes(eof_ts_std
       (ieof- 1,:)),1),eof_ts_std(ieof- 1,:),xyres)
68   end do
69   resP= True
70   resP@ gsnFrame= False
71   gsn_panel(wks,plot, (/4,2/),resP)
72   frame(wks)
73   end
```

Line 1 is the beginning of program.

Line 2—3 open the file and read in data.

Line 4 sets the number of EOFs.

Line 5 computes empirical orthogonal functions.

Line 7 calculates the time series of the amplitudes associated with each eigenvalue in an EOF.

Line 8 standardizes the time series.

Line 14—40 set the parameters associated with EOF patterns.

Line 41—60 set the parameters associated with EOF time series.

Line 61—65 plot EOF patterns.

Line 66—67 plot EOF time series.

Line 69—71 create panel plot.

Results:

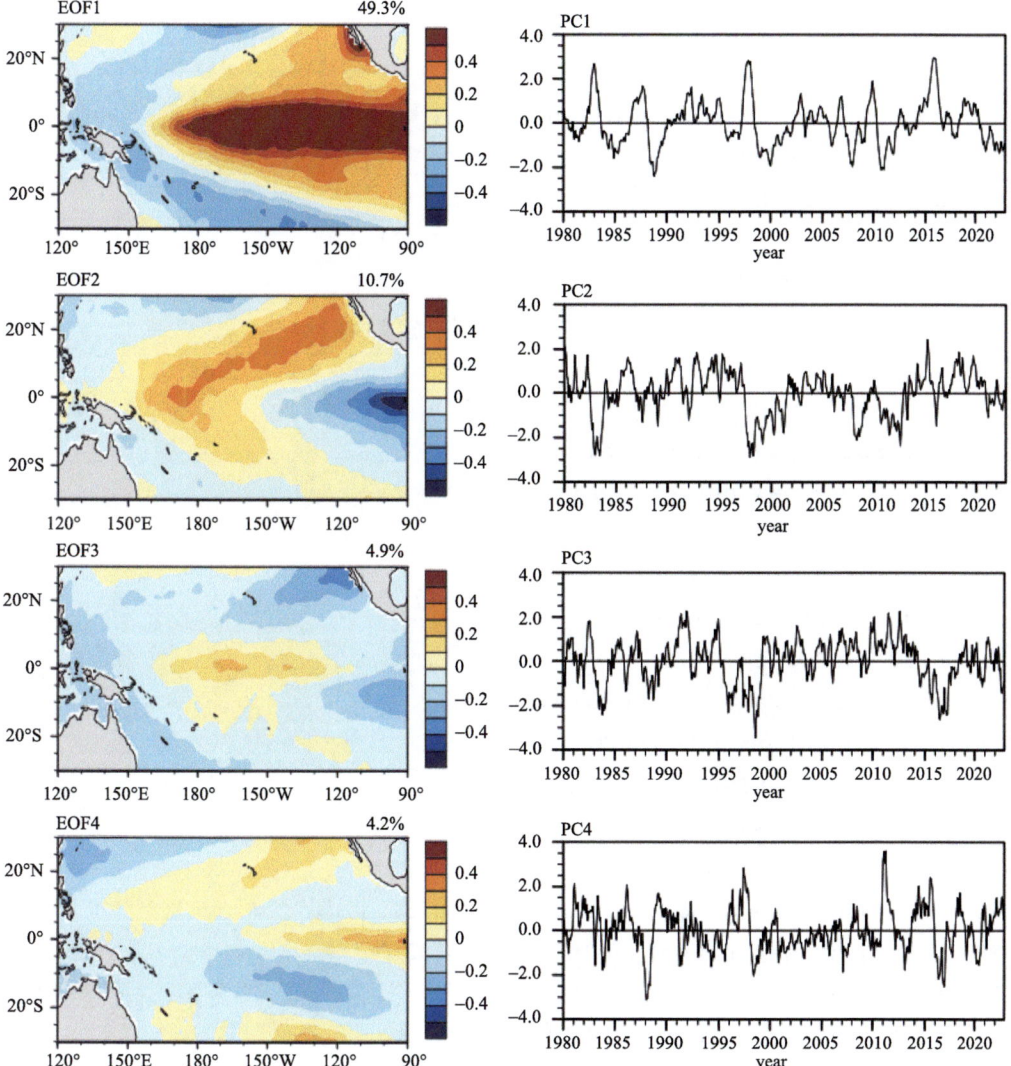

Example 8-2-2:

Calculate the first two EOFs of sea level pressure (SLP) north of 20°N. The first EOF pattern is commonly identified as the Arctic Oscillation (AO) mode. The data used in the following example comes from this web site: https://psl.noaa.gov/data/gridded/data.ncep.reanalysis2.html.

Code:

```
1    begin
2    f= addfile ("./pres_NCEP2_198001- 202012.nc","r")
3    slp= f- > pres(::12,{20:90},:)
4    w= sqrt(cos(0.01745329* slp&lat))
5    wp= slp* conform(slp, w, 1)
6    copy_VarCoords(slp, wp)
7    x= wp(lat|:,lon|:,time|:)
8    neof= 2
9    eof= eofunc_Wrap(x, neof, False)
10   eof= eof* - 1.0
11   eof_ts= eofunc_ts_Wrap(x, eof, False)
12   eof_ts= dim_standardize_n(eof_ts, 1, 1)
13   do n= 0,neof- 1
14   eof(n,:,:)= eof(n,:,:) * stddev(eof_ts(n,:))
15   end do
16   wks = gsn_open_wks("pdf","ex8.2.2")
17   plotEOF= new(neof,graphic)
18   plottime= new(neof,graphic)
19   res= True
20   res@ gsnDraw= False
21   res@ gsnFrame= False
22   res@ gsnPolar= "NH"
23   res@ cnFillPalette= "BlueDarkRed18"
24   res@ mpCenterLonF= 180.0
25   res@ mpMinLatF= 20
26   res@ cnFillOn= True
27   res@ lbLabelBarOn= False
28   res@ gsnLeftStringFontHeightF= 0.035
29   res@ gsnRightStringFontHeightF= 0.035
30   do n= 0,neof- 1
31   res@ gsnLeftString= "EOF "+ (n+ 1)
32   res@ gsnRightString= sprintf("% 5.1f", eof@ pcvar(n)) + "% "
33   plotEOF(n)= gsn_csm_contour_map_polar(wks,eof(n,:,:),res)
```

```
34  end do
35  resP= True
36  resP@ gsnMaximize= True
37  resP@ gsnPanelXWhiteSpacePercent= 3
38  resP@ gsnPanelLabelBar= True
39  gsn_panel(wks,plotEOF,(/2,1/),resP)
40  rts= True
41  rts@ gsnDraw= False
42  rts@ gsnFrame= False
43  rts@ tmXTOn= False
44  rts@ tmYROn= False
45  rts@ vpHeightF= 0.3
46  rts@ vpWidthF= 0.6
47  rts@ gsnYRefLine= 0.0
48  rts@ gsnAboveYRefLineColor= "red"
49  rts@ gsnBelowYRefLineColor= "blue"
50  rts@ tiYAxisString= ""
51  rts@ xyLineThicknesses = 2.0
52  rts@ tmXBMode= "Explicit"
53  rts@ tmXBValues= (/0,5,10,15,20,25,30,35,40/)
54  rts @ tmXBLabels = (/"1980","1985","1990","1995","2000","2005",
     "2010","2015","2020"/)
55  rts@ tmXBLabelAngleF = 30
56  rts@ trYMaxF= 3
57  rts@ trYMinF= - 3
58  do n= 0,neof- 1
59  rts@ gsnLeftString= "PC"+ (n+ 1)
60  rts@ gsnRightString= sprintf("% 5.1f", eof@ pcvar(n)) + "% "
61  plottime(n)= gsn_csm_xy (wks,ispan(0, 40, 1),eof_ts(n,:),rts)
62  end do
63  rtsP= True
64  gsn_panel(wks,plottime,(/2,1/),rtsP)
65  end
```

Line 2—3 open the file and read the Northern Hemisphere data for January each year.

Line 4—6 calculate the weights.

Line 7 indicates that the time dimension is on the far right.

Line 8 keeps the first two EOF modes.

Line 9 computes empirical orthogonal functions.

Line 11 calculates the time series of the amplitudes associated with each eigenvalue in the EOF.

Line 12 standardizes the time series.

Line 17—18 create graphic arrays.

Line 19—39 set the parameters associated with EOF patterns and plot EOF patterns.

Line 40—64 set the parameters associated with EOF time series and plot EOF time series.

Results:

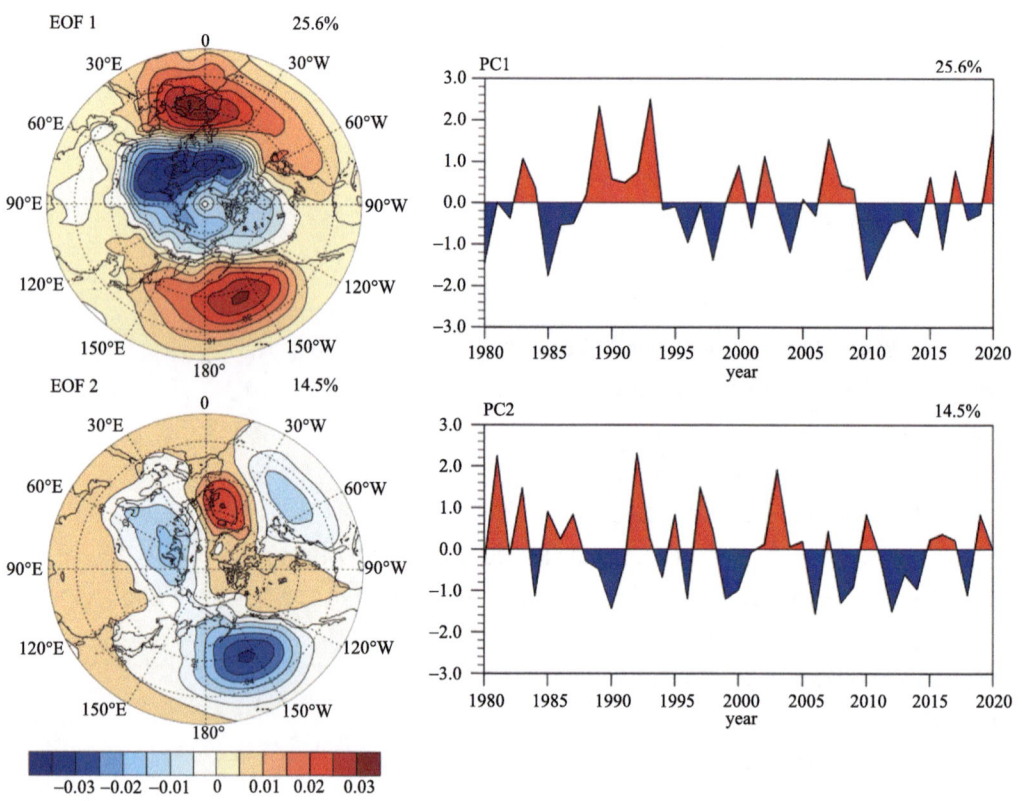

Example 8-2-3:

Calculate the first two EOFs of sea level pressure (SLP) over the North Pacific. The first EOF pattern is commonly identified as the Aleutian low (AL) mode. The second EOF mode is commonly identified as the North Pacific Oscillation (NPO). The data is the same as that in Example 8-2-2.

Code:

```
1    Begin
2    f= addfile ("./pres_NCEP2_198001- 202012.nc","r")
3    slp= f- > pres(:,{20:70},{120:280})
4    w= sqrt(cos(0.01745329* slp&lat))
5    wp= slp* conform(slp, w, 1)
6    copy_VarCoords(slp, wp)
7    x= wp(lat|:,lon|:,time|:)
8    neof= 2
```

```
9   eof= eofunc_Wrap(x, neof, False)
10  eof= eof* - 1.0
11  eof_ts= eofunc_ts_Wrap(x, eof, False)
12  eof_ts= dim_standardize_n(eof_ts, 1, 1)
13  do n= 0,neof- 1
14  eof(n,:,:)= eof(n,:,:)* stddev(eof_ts(n,:))
15  end do
16  wks = gsn_open_wks("pdf","ex8.2.3")
17  plotEOF= new(neof,graphic)
18  plottime= new(neof,graphic)
19  res= True
20  res@ gsnDraw= False
21  res@ gsnFrame= False
22  res@ mpShapeMode= "FreeAspect"
23  res@ vpWidthF= 0.7
24  res@ vpHeightF= 0.4
25  res@ gsnAddCyclic= False
26  res@ cnFillPalette= "BlueDarkRed18"
27  res@ mpCenterLonF= 180.0
28  res@ mpMinLatF= 20
29  res@ mpMaxLatF= 70
30  res@ mpMinLonF= 120
31  res@ mpMaxLonF= 280
32  res@ cnFillOn= True
33  res@ lbLabelBarOn= False
34  do n= 0,neof- 1
35  res@ gsnLeftString= "EOF "+ (n+ 1)
36  res@ gsnRightString= sprintf("% 5.1f", eof@ pcvar(n)) + "% "
37  plotEOF(n)= gsn_csm_contour_map_polar(wks,eof(n,:,:),res)
38  end do
39  resP= True
40  resP@ gsnMaximize= True
41  resP@ gsnPanelXWhiteSpacePercent= 3
42  resP@ gsnPanelLabelBar= True
43  gsn_panel(wks,plotEOF,(/2,1/),resP)
44  rts= True
45  rts@ gsnDraw= False
46  rts@ gsnFrame= False
47  rts@ tmXTOn= False
```

```
48  rts@ tmYROn= False
49  rts@ vpHeightF= 0.7
50  rts@ vpWidthF= 0.7
51  rts@ gsnYRefLine= 0.0
52  rts@ gsnAboveYRefLineColor= "red"
53  rts@ gsnBelowYRefLineColor= "blue"
54  rts@ tiYAxisString= ""
55  rts@ xyLineThicknesses= 2.0
56  rts@ tmXBMode= "Explicit"
57  rts@ tmXBValues= (/0,60,120,180,240,300,360,420,480/)
58   rts @ tmXBLabels = (/"198001","198501","199001","199501","200001",
       "200501","201001","201501","202001"/)
59  rts@ tmXBLabelAngleF = 30
60  rts@ trYMaxF= 3
61  rts@ trYMinF= -3
62  do n= 0,neof-1
63  rts@ gsnLeftString= "PC"+ (n+1)
64  rts@ gsnRightString= sprintf("% 5.1f", eof@ pcvar(n)) + "% "
65  plottime(n)= gsn_csm_xy (wks,ispan(0, 40, 1),eof_ts(n,:),rts)
66  end do
67  rtsP= True
68  gsn_panel(wks,plottime,(/2,1/),rtsP)
69  end
```

Line 2—3 open the file and read the data of 20°N—70°N.

Line 4—6 calculate the weights.

Line 7 indicates that the time dimension is on the far right.

Line 8 keeps the first two EOF modes.

Line 9 computes empirical orthogonal functions.

Line 11 calculates the time series of the amplitudes associated with each eigenvalue in an EOF.

Line 12 standardizes the time series.

Line 17—18 create graphic arrays.

Line 19—43 set the parameters associated with EOF patterns and plot EOF patterns.

Line 44—68 set the parameters associated with EOF time series and plot EOF time series.

Results:

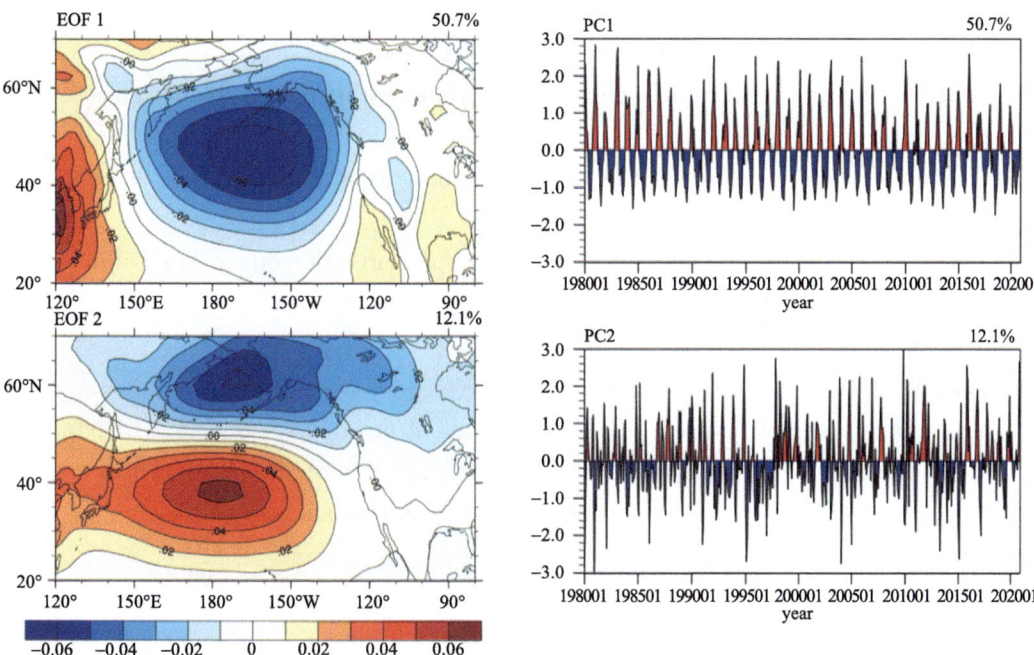

➢ **Practices**

The sea level pressure data used in the following question comes from this web site: https://psl.noaa.gov/data/gridded/data.ncep.reanalysis2.html.

The sea surface temperature data used in the following question comes from this web site: https://www.metoffice.gov.uk/hadobs/hadisst/data/download.html.

(1) Read sea level pressure; extract the data spanning 1979−2021; calculate the first three EOFs over the North Atlantic region for the summer (JJA) season.

(2) Read sea surface temperature; extract the data spanning 1979−2021; calculate the first two EOFs over the South Pacific region.

8.3 Composite analysis

Composite Analysis is a statistical technique used to study the average behavior of variables of interest under different conditions or states. It involves dividing the data into subsets based on some criteria, such as time periods or regions, and then computing composite averages for each subset. This method is useful for identifying patterns in complex datasets and can provide insights into how different factors may be contributing to observed variations. Composite analysis has applications in several fields including meteorology, climate science, ecology, and economics.

➢ **Key points**

Composite functions:

$$\text{dim\_avg\_n\_Wrap}(x, \text{dims})$$

The arguments used in the function are set as follows:

• x: A variable of numeric type and any dimensionality.

• $dims$: The dimension(s) of x on which to calculate the average. Must be consecutive and monotonically increasing.

• This function is identical to **dim_avg_n**, except the return value will have metadata added based on metadata attached to x.

• dim: The dimension index of data that represents the dimension containing the number of observations. Generally, this is the time dimension.

➢ **Examples**

Example 8-3-1:

The composite analysis of the sea surface temperature in December for El Niño years during 1979—2016. The SST data used in the following example comes from this web site: https://www.metoffice.gov.uk/hadobs/hadisst/data/download.html.

Code:

```
1   Begin
2   f= addfile("./ersst.v5.1979-2016.nc","r")
3   sst= f->sst(11::12,0,{-60:60},{120:270})
4   ssta= dim_rmvmean_n_Wrap(sst,0)
5   ssta&time= ispan(1979,2016,1)
6   S= ssta&time
7   dim= ind(S.eq.1982.or.S.eq.1986.or.S.eq.1987.or.S.eq.1991.or.S.eq.1994.or.S.eq.1997.or.S.eq.2002.or.S.eq.2004.or.S.eq.2006.or.S.eq.2009.or.S.eq.2014.or.S.eq.2015)
8   newssta= ssta(dim,:,:)
9   elnino_year= dim_avg_n_Wrap(newssta,0)
10  wks= gsn_open_wks("pdf","ex8.3.1")
11  res= True
12  res@ gsnDraw= False
13  res@ gsnFrame= False
14  res@ gsnAddCyclic= False
15  res@ tmXTOn= False
16  res@ tmYROn= False
17  res@ mpCenterLonF= 180.0
18  res@ mpMinLatF= -60.0
19  res@ mpMaxLatF= 60.0
20  res@ mpMinLonF= 120
21  res@ mpMaxLonF= 270
22  res@ mpFillOn= False
```

```
23    res@ cnFillOn= True
24    res@ cnFillPalette = "BlueDarkRed18"
25    res@ cnLinesOn= False
26    res@ cnLineLabelsOn= False
27    res@ lbLabelBarOn= True
28    res@ gsnLeftString = ""
29    res@ gsnRightString= ""
30    plot= gsn_csm_contour_map(wks, elnino_year, res)
31    draw(plot)
32  end
```

Line 1 is the beginning of program.

Line 2—3 open the file and read in data.

Line 4 calculates the SST anomaly.

Line 5—8 select El Niño years.

Line 9 is the composite of El Niño years.

Line 11—29 set the parameters associated with the composite plot of El Niño years.

Line 30 creates the composite plot.

Results:

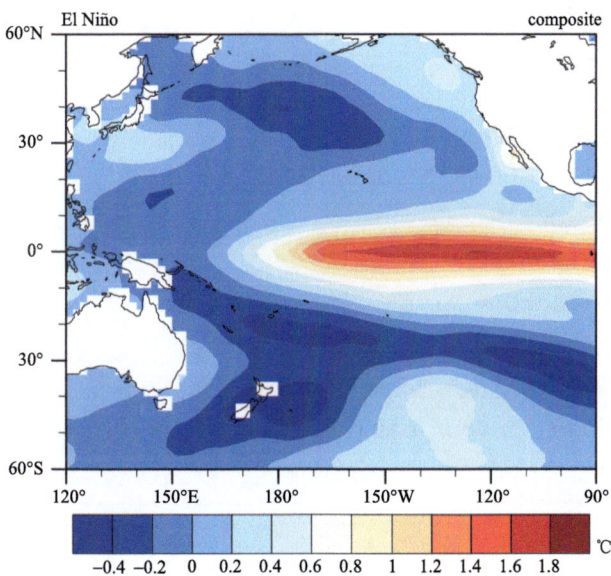

Example 8-3-2:

The composite analysis of the sea surface temperature in December for La Niña years during 1979—2016. The SST data is the same as that in Example 8-3-1.

Code:

```
1  begin
2    f= addfile("./ersst.v5.1979- 2016.nc","r")
3    sst= f- > sst(11::12,0,{- 60:60},{120:270})
```

```
4   ssta= dim_rmvmean_n_Wrap(sst, 0)
5   ssta&time= ispan(1979, 2016, 1)
6   S= ssta&time
7   dim= ind(S.eq.1983.or.S.eq.1984.or.S.eq.1988.or.S.eq.1995.or.S.
    eq.1998.or.S.eq.1999.or.S.eq.2000.or.S.eq.2005.or.S.eq.2007.or.S.
    eq.2008.or.S.eq.2010.or.S.eq.2011.or.S.eq.2016)
8   newssta= ssta(dim,:,:)
9   lanina_year= dim_avg_n_Wrap(newssta,0)
10  wks= gsn_open_wks("pdf","ex8.3.2")
11  res= True
12  res@ gsnDraw= False
13  res@ gsnFrame= False
14  res@ gsnAddCyclic= False
15  res@ tmXTOn= False
16  res@ tmYROn= False
17  res@ mpCenterLonF= 180.0
18  res@ mpMinLatF= - 60.0
19  res@ mpMaxLatF= 60.0
20  res@ mpMinLonF= 120
21  res@ mpMaxLonF= 270
22  res@ mpFillOn= False
23  res@ cnFillOn= True
24  res@ cnFillPalette = "BlueDarkRed18"
25  res@ cnLinesOn= False
26  res@ cnLineLabelsOn= False
27  res@ lbLabelBarOn= True
28  res@ gsnLeftString = "La Nina"
29  res@ gsnRightString= "composite "
30  plot= gsn_csm_contour_map(wks, lanina_year, res)
31  draw(plot)
32  end
```

Line 5—8 select La Niña years.

Line 9 is the composite of La Niña years.

Results:

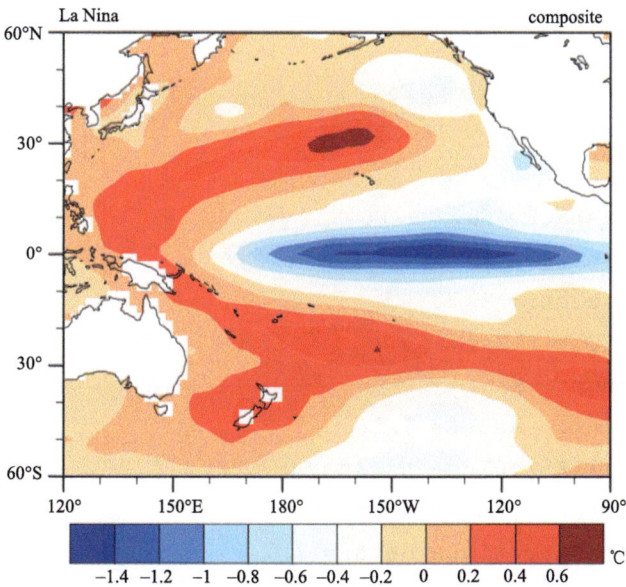

Example 8-3-3:

The composite analysis of the horizontal winds at 850 hPa in December for El Niño years during 1979−2020. The data used in the following example comes from this web site: https://psl. noaa. gov/data/gridded/data. ncep. reanalysis2. html.

Code:

```
1    begin
2    fu= addfile("./uwnd.mon.mean197901- 202104.nc","r")
3    u= fu- > uwnd(11::12,2,{- 60:60},{120:270})
4    fv= addfile("./vwnd.mon.mean197901- 202104.nc","r")
5    v= fv- > vwnd(11::12,2,{- 60:60},{120:270})
6    ua= dim_rmvmean_n_Wrap(u, 0)
7    va= dim_rmvmean_n_Wrap(v, 0)
8    ua&time= ispan(1979, 2020, 1)
9    va&time= ispan(1979, 2020, 1)
10   S= ua&time
11   dim= ind(S.eq.1982.or.S.eq.1986.or.S.eq.1987.or.S.
     eq.1994.or.S.eq.1997.or.S.eq.2002.or.S.eq.2004.or.S.
     eq.2009.or.S.eq.2014.or.S.eq.2015.or.S.eq.2018)
12   new_ua= ua(dim,:,:)
13   new_va= va(dim,:,:)
14   ua_elnino_year= dim_avg_n_Wrap(new_ua,0)
15   va_elnino_year= dim_avg_n_Wrap(new_va,0)
16   wks= gsn_open_wks("pdf","ex8.3.3")
```

```
17    resv= True
18    resv@ gsnDraw= False
19    resv@ gsnFrame= False
20    resv@ gsnAddCyclic= False
21    resv@ mpCenterLonF= 180.0
22    resv@ mpMinLatF= -60.0
23    resv@ mpMaxLatF= 60.0
24    resv@ mpMinLonF= 120
25    resv@ mpMaxLonF= 270
26    resv@ tmXTOn= False
27    resv@ tmYROn= False
28    resv@ vcMinDistanceF= 0.02
29    resv@ vcFillArrowEdgeThicknessF= 1.2
30    resv@ vcGlyphStyle= "CurlyVector"
31    resv@ vcLineArrowThicknessF= 2.0
32    resv@ vcRefAnnoOn= True
33    resv@ vcRefMagnitudeF= 5.0
34    resv@ vcRefAnnoString1On= True
35    resv@ vcRefAnnoString1= "5.0m/s"
36    resv@ vcRefAnnoSide= "Top"
37    resv@ vcRefAnnoString2On= False
38    resv@ vcRefAnnoPerimOn= True
39    resv@ vcRefAnnoOrthogonalPosF= -0.17
40    resv@ vcRefAnnoParallelPosF= 1
41    resv@ vcRefLengthF= 0.1
42    resv@ vcRefAnnoBackgroundColor= "White"
43    resv@ vcRefAnnoFontHeightF= 0.02
44    plot= gsn_csm_vector_map(wks, ua_elnino_year, va_elnino_year, resv)
45    draw(plot)
46    end
```

Line 2—5 open the file and read in data.

Line 6—7 calculate the wind anomaly.

Line 8—13 select El Niño years.

Line 14—15 is the composite of El Niño years.

Line 17—43 set the vector field parameters associated with the composite plot of El Niño years.

Line 44 creates the composite plot.

Results:

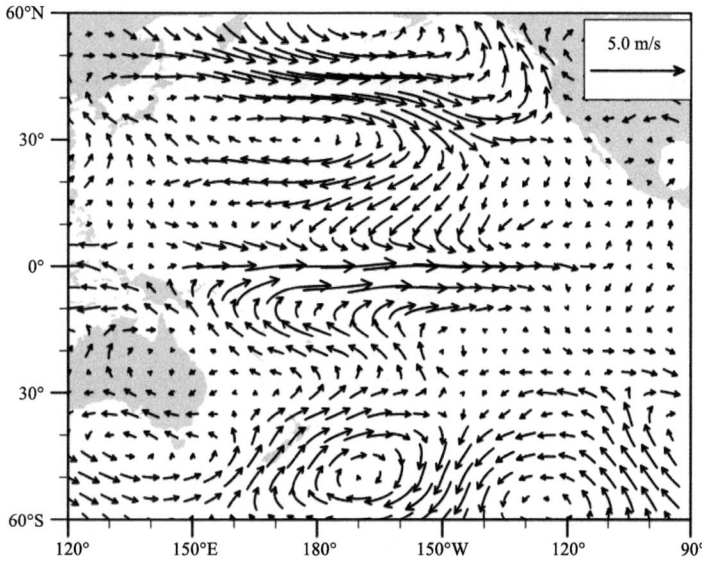

> **Practices**

The data used in the following questions comes from this web site: https://psl.noaa.gov/data/gridded/data.ncep.reanalysis2.html.

(1) Draw composite plot of horizontal winds at 850 hPa in winter for La Niña years during 1979—2020.

(2) Draw composite plot of geopotential height field at 500 hPa in summer during 1979—2020.

8.4 Regression and correlation analysis

Regression and correlation analysis are statistical methods used to investigate the relationship between two or more variables. Regression analysis tests the linear relationship between a dependent variable (y) and one or more independent variables (x). The aim is to develop a mathematical equation that can predict the value of y based on x. Correlation analysis, on the other hand, tests how well two variables are related to each other without determining any cause-and-effect relationship between them.

> **Key points**

■ Correlation analysis

(1) Compute sample cross-correlations

```
esccr(x,y,mxlag)
```

The arguments used in the function are set as follows:

- x: An array of any numeric type or size. The rightmost dimension is usually time.
- y: An array of any numeric type or size. The rightmost dimension is usually time.

The size of the rightmost dimension must be the same as x.

• $mxlag$: A scalar integer. It is recommended that $0 \leqslant mxlag \leqslant N/4$. This is because the correlation algorithm(s) use N rather than $(N-k)$ values in the denominator.

(2) Compute the (Pearson) sample linear cross-correlations at lag 0 only

```
escorc(x,y)
```

The arguments used in the function are set as follows:

• x: An array of any numeric type or size. The rightmost dimension is usually time.

• y: An array of any numeric type or size. The rightmost dimension is usually time. The size of the rightmost dimension must be the same as x.

(3) Compute sample auto-covariances

```
esacv(x,maxlag)
```

The arguments used in the function are set as follows:

• x: An array of any numeric type or size. The rightmost dimension is usually time.

• $mxlag$: A scalar integer. It is recommended that $0 \leqslant mxlag \leqslant N/4$. This is because the correlation algorithm(s) use N rather than $(N-k)$ values in the denominator.

(4) Compute sample auto-correlations

```
esacr(x,maxlag)
```

The arguments used in the function are the same as for the `esacv`.

(5) Compute sample cross-covariances

```
esccv(x,y,maxlag)
```

The arguments used in the function are the same as for the `esccr`.

(6) Compute sample cross-covariances at lag 0 only

```
escovc(x,y)
```

The arguments used in the function are the same as for the `escorc`.

■ Regression analysis

(1) Calculate the linear regression coefficient between two series

```
regline(x,y)
```

The arguments used in the function are set as follows:

• x/y: One-dimensional arrays of the same length.

(2) Calculate the linear regression coefficient on multi-dimensional arrays

```
regCoef(x,y)
```

The arguments used in the function are set as follows:

• x: An array of any dimensionality.

• y: An array of any dimensionality. The last (rightmoast) dimension of y must be the same as the last dimension of x.

(3) Calculate the linear regression coefficient between two variables on the given dimensions

```
regCoef_n(x,y,dims_x,dims_y)
```

The arguments used in the function are set as follows:

- x/y: An array of any dimensionality.
- $dims\_x/y$: A scalar integer indicating which dimension of x/y to do the calculation on.

➢ **Examples**

Example 8-4-1:

Calculate the cross correlation between two variables, and calculate positive and negative lags in a cross correlation. The SST data used in the following example comes from this web site: https://www.metoffice.gov.uk/hadobs/hadisst/data/download.html. The SLP data comes from this web site: https://psl.noaa.gov/data/gridded/data.ncep.reanalysis2.html.

Code:

```
1    begin
2    fs= addfile("./sst_ERSSTv5_198001_202012_grid1.nc","r")
3    sst= fs-> sst(:,{-5:5},{190:240})
4    sst_avg= clmMonTLL(sst)
5    ssta= calcMonAnomTLL(sst, sst_avg)
6    nino3.4= wgt_areaave_Wrap(ssta, 1.0, 1.0, 1)
7    fp= addfile ("./pres_NCEP2_198001- 202012_grid1.nc","r")
8    slp= fp-> pres(:,{30:65},{160:220})
9    SLP_avg= clmMonTLL(slp)
10   SLPA= calcMonAnomTLL(slp, SLP_avg)
11   ALI= wgt_areaave_Wrap(SLPA, 1.0, 1.0, 1)
12   maxlag= 0
13   cor0= esccr(nino3.4,ALI,maxlag)
14   maxlag= 12
15   ALI_lead_nino3.4= esccr(ALI,nino3.4,maxlag)
16   nino3.4_lead_ALI= esccr(nino3.4,ALI,maxlag)
17   cor= new ( 2* maxlag+ 1, float)
18   cor(0:maxlag)= nino3.4_lead_ALI (::- 1)
19   cor(maxlag+ 1:)= ALI_lead_nino3.4(1:)
20   x= ispan(- maxlag,maxlag,1)
21   wks= gsn_open_wks("pdf","ex8.4.1")
22   res= True
23   res@ tiMainString= "Cor(nino3.4 vs ALI)"
24   res@ tiXAxisString= "LAG(month)"
25   res@ xyLineThicknessF= 4.0
26   plot= gsn_xy(wks,x,cor,res)
27   end
```

Line 1 is the beginning of program.

Line 2—3 open the file and read SST data.
Line 4—5 calculate the SST anomaly.
Line 6 calculates the Niño3.4 index.
Line 7—8 open the file and read SLP data.
Line 9—10 calculate the SLP anomaly.
Line 11 calculates the Aleutian low index.
Line 12 sets lag.
Line 13 calculates the simultaneous correlation coefficient.
Line 14 sets lag.
Line 15 calculates positive lag cross correlation.
Line 16 calculates negative lag cross correlation.
Line 17 sets total lag and allocates memory.
Line 18 indicates that：：—1 means to reverse the order.
Line 20 defines X-axis.
Line 21—26 plot the correlations.

Results:

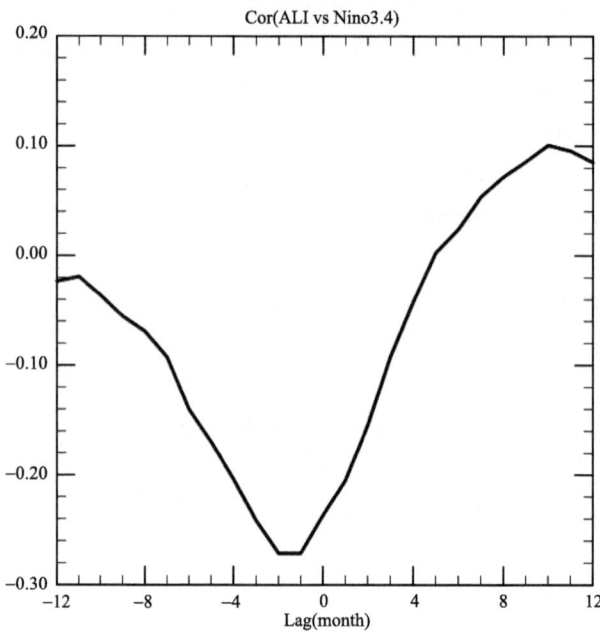

Example 8-4-2:

Calculate a two-dimensional correlation at lag 0. The data is the same as that in Example 8-4-1.

Code (way 1):

```
1   begin
2   fp= addfile ("./pres_NCEP2_198001- 202012_grid1.nc","r")
3   slp= fp- > pres
```

```
4    SLP_avg= clmMonTLL(slp)
5    slpa= calcMonAnomTLL(slp, SLP_avg)
6    fs= addfile("./sst_ERSSTv5_198001_202012_grid1.nc","r")
7    sst= fs- > sst
8    sst_avg= clmMonTLL(sst)
9    ssta= calcMonAnomTLL(sst, sst_avg)
10   slpa:= slpa(lat|:,lon|:,time|:)
11   ssta:= ssta(lat|:,lon|:,time|:)
12   maxlag= 2
13   cor= esccr(slpa,ssta,maxlag)
14   copy_VarAtts(slpa, cor)
15   copy_VarCoords_1(slpa,cor)
16   wks= gsn_open_wks("pdf","ex8.4.2.1")
17   res= True
18   res@ cnFillOn= True
19   res@ cnFillPalette= "BlWhRe"
20   res@ cnLinesOn= False
21   res@ cnLevelSelectionMode= "ManualLevels"
22   res@ cnMinLevelValF= - 1.0
23   res@ cnMaxLevelValF = 1.0
24   res@ cnLevelSpacingF = 0.1
25   res@ pmLabelBarWidthF= 0.8
26   lag= 0
27   res@ tiMainString= "Correlations at lag "+ lag
28   plot= gsn_csm_contour_map(wks,cor(:,:,lag),res)
29   end
```

Line 1 is the beginning of program.

Line 2—3 open the file and read SLP data.

Line 4—5 calculate the SLP anomaly.

Line 6—7 open the file and read SST data.

Line 8—9 calculate the SST anomaly.

Line 10—11 reorder to get time as the right most dimension.

Line 12 sets lag.

Line 13 calculates cross correlations.

Line 14—15 copy meta data and coordinate variables using contributed functions.

Line 16—28 set the drawing parameters and draw.

Results:

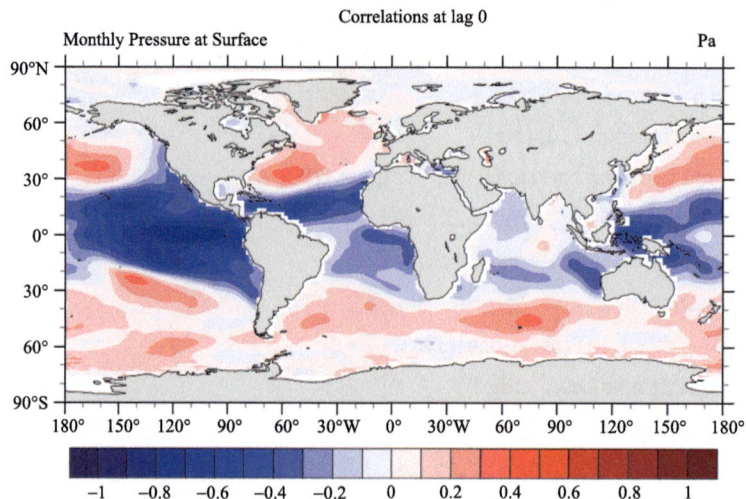

Code (way 2):

```
1   begin
2   fp= addfile ("./pres_NCEP2_198001- 202012_grid1.nc","r")
3   slp= fp- > pres
4   SLP_avg= clmMonTLL(slp)
5   slpa= calcMonAnomTLL(slp, SLP_avg)
6   fs= addfile("./sst_ERSSTv5_198001_202012_grid1.nc","r")
7   sst= fs- > sst
8   sst_avg= clmMonTLL(sst)
9   ssta= calcMonAnomTLL(sst, sst_avg)
10  slpa:= slpa(lat|:,lon|:,time|:)
11  ssta:= ssta(lat|:,lon|:,time|:)
12  cor= escorc(slpa,ssta)
13  copy_VarAtts(slpa, cor)
14  copy_VarCoords_1(slpa,cor)
15  wks= gsn_open_wks("pdf","ex8.4.2.2")
16  res= True
17  res@ cnFillOn= True
18  res@ cnFillPalette= "BlWhRe"
19  res@ cnLinesOn= False
20  res@ cnLevelSelectionMode= "ManualLevels"
21  res@ cnMinLevelValF= - 1.0
22  res@ cnMaxLevelValF= 1.0
23  res@ cnLevelSpacingF = 0.1
24  res@ pmLabelBarWidthF= 0.8
```

```
25    res@ tiMainString= "Correlations at lag 0"
26    plot= gsn_csm_contour_map(wks,cor,res)
27    end
```
Line 12 computes the correlation at each latitude and longitude at lag 0.

Results:

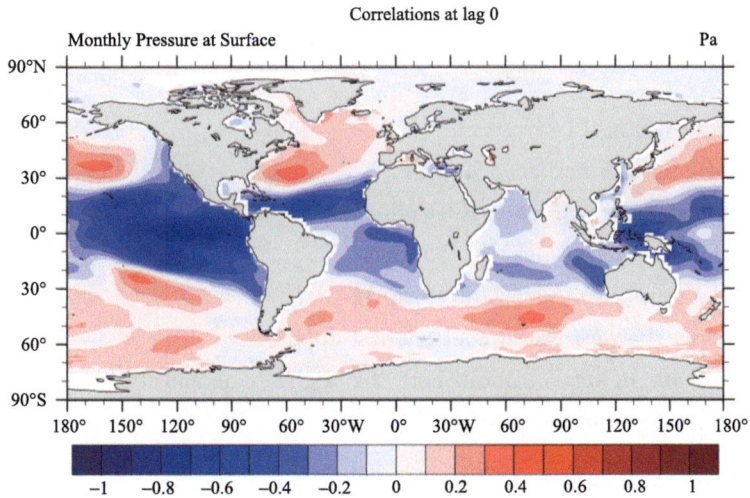

Example 8-4-3:

Calculate the least squared regression for a one-dimensional array. The data used in the following example comes from this web site: https://www.ncl.ucar.edu/Applications/Data.

Code:
```
1     begin
2     ncol= 2
3     ntim= numAsciiRow("./regress_1.txt")
4     data= asciiread("./regress_1.txt", (/ntim,ncol/), "float")
5     data@ _FillValue= - 9999.0
6     x= data(:,0)
7     y= data(:,1)
8     rc= regline(x, y)
9     print(rc)
10    rc@ units= "degK/day"
11    pltarry= new((/2,ntim/), typeof(data), data@ _FillValue)
12    pltarry(0,:)= y
13    pltarry(1,:)= rc* x + rc@ yintercept
14    wks= gsn_open_wks("pdf","ex8.4.3")
15    res= True
16    res@ xyMarkLineModes= (/"Markers","Lines"/)
```

```
17  res@ xyMarkers= 16
18  res@ xyMarkerColor= "red"
19  res@ xyMarkerSizeF= 0.005
20  res@ xyDashPatterns= 1
21  res@ xyLineThicknesses= (/1,2/)
22  res@ tiMainString= "Output from regline"
23  plot= gsn_csm_xy (wks,x,pltarry,res)
24  end
```

Line 1 is the beginning of program.

Line 2—4 open the file and read in data.

Line 5 sets the missing values (_FillValue) to −9999.0.

Line 6 sets the units of model time to days.

Line 7 sets the units of model value to degK.

Line 8 calculates the regression coefficient (slope).

Line 11—13 create an array to hold both the original data and the calculated regression line.

Line 14 converts graphics to pdf file.

Line 15 sets the desired plot mods.

Line 16 chooses which have markers.

Line 17 chooses the type of marker.

Line 18 sets the marker color.

Line 19 sets the marker size.

Line 20 sets solid line.

Line 21 sets the second line to 2.

Line 22 sets the title.

Line 23 creates the plot.

Results:

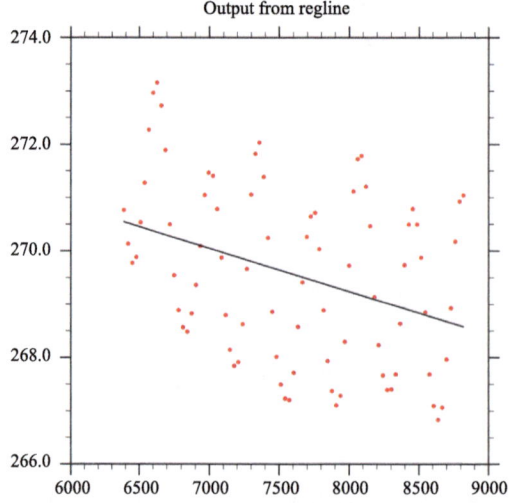

Example 8-4-4:

Calculate the regression coefficient at each grid point. The SST data comes from this web site: https://www.metoffice.gov.uk/hadobs/hadisst/data/download.html. The SLP data comes from this web site: https://psl.noaa.gov/data/gridded/data.ncep.reanalysis2.html.

Code:

```
1   begin
2   fS= addfile("./ersst.v5.1950_2018.nc","r")
3   sst= fS-> sst(374:733:12,0,{-30:60},{120:270})
4   ssta= dim_rmvmean_n_Wrap(sst,0)
5   f= addfile("./pres_NCEP2_198001-202012.nc","r")
6   slp= f-> pres(12:371,{20:60},{120:280})
7   SLP_avg= clmMonTLL(slp)
8   SLPA= calcMonAnomTLL(slp, SLP_avg)
9   w= sqrt(cos(0.01745329* SLPA&lat))
10  wp= SLPA* conform(SLPA, w, 1)
11  copy_VarCoords(SLPA, wp)
12  x= wp(lat|:,lon|:,time|:)
13  neof= 2
14  eof= eofunc_Wrap(x, neof, False)
15  eof_ts= eofunc_ts_Wrap(x, eof, False)
16  eof_ts:= dim_standardize_n(eof_ts, 1, 1)
17  eof_ts(1,:)= -eof_ts(1,:)
18  AL= eof_ts(0,2::12)
19  reg= regCoef(AL, ssta(lat|:,lon|:,time|:))
20  reg!0= "lat"
21  reg!1= "lon"
22  reg&lat= ssta&lat
23  reg&lon= ssta&lon
24  wks= gsn_open_wks("pdf","ex8.4.4")
25  res= True
26  res@ gsnDraw= False
27  res@ gsnFrame= False
28  res@ gsnAddCyclic= False
29  res@ mpFillOn= False
30  res@ mpCenterLonF= 180.0
31  res@ mpMinLatF= -30.0
32  res@ mpMaxLatF= 60.0
33  res@ mpMinLonF= 120.0
```

```
34    res@ mpMaxLonF= 270.0
35    res@ cnFillOn= True
36    res@ cnLinesOn= False
37    res@ cnFillPalette = "amwg_blueyellowred"
38    res@ gsnSpreadColorStart= 2
39    res@ gsnSpreadColorEnd= 13
40    res@ cnLevelSelectionMode= "ManualLevels"
41    res@ cnMinLevelValF= - 0.4
42    res@ cnMaxLevelValF= 0.4
43    res@ cnLevelSpacingF= 0.1
44    res@ tiMainString= "ssta vs AL_PC1"
45    plot= gsn_csm_contour_map(wks, reg, res)
46    draw(plot)
47    end
```

Line 1 is the beginning of program.

Line 2—3 open the file and read the March SST data during 1981—2010.

Line 4 calculates the SST anomaly.

Line 5—6 open the file and read the March SLP during 1981—2010.

Line 7—8 calculate the SLP anomaly.

Line 9—10 calculate the weights.

Line 12 the time dimension is on the far right.

Line 13 keeps the first two EOF modes.

Line 14 computes empirical orthogonal functions.

Line 15 calculates the time series of the amplitudes associated with each eigenvalue in an EOF.

Line 16 standardizes the time series.

Line 18 selects the March PC1 as the Aleutian low index.

Line 19 calculates the regression coefficient.

Line 25—44 set plotting parameters.

Line 45—46: create the plot.

Results:

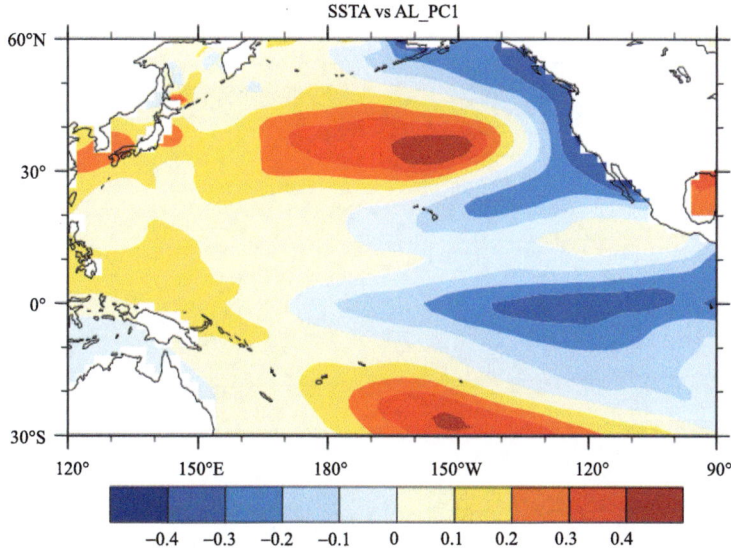

> **Practices**

The ENSO index comes from this web site: https://climatedataguide.ucar.edu/climate-data/nino-sst-indices-nino-12-3-34-4-oni-and-tni.

The sst data is derived from https://www.metoffice.gov.uk/hadobs/hadisst/data/download.html.

The Arctic Oscillation (AO) index is derived from http://www.cpc.ncep.noaa.gov.

The SLP data and wind field data are derived from https://psl.noaa.gov/data/gridded/data.ncep.reanalysis2.html.

The Aleutian low index is defined as area-averaged SLP anomalies over 30°N—65°N, 160°E—140°W.

(1) Calculate the correlation coefficient between Arctic Oscillation index and ENSO index.

(2) Calculate the correlation between the Aleutian low index and SST field at lags 5.

(3) Calculate the regression coefficient at each grid point about regression of 850 hPa winds in winter onto the March Aleutian low index.